Júlio Nonato Silva Nascimento

Recovering degraded areas under the influence of urban waste

Júlio Nonato Silva Nascimento

Recovering degraded areas under the influence of urban waste

Recovering springs in the Amazon

ScienciaScripts

Imprint
Any brand names and product names mentioned in this book are subject to trademark, brand or patent protection and are trademarks or registered trademarks of their respective holders. The use of brand names, product names, common names, trade names, product descriptions etc. even without a particular marking in this work is in no way to be construed to mean that such names may be regarded as unrestricted in respect of trademark and brand protection legislation and could thus be used by anyone.

Cover image: www.ingimage.com

This book is a translation from the original published under ISBN 978-3-330-75598-7.

Publisher:
Sciencia Scripts
is a trademark of
Dodo Books Indian Ocean Ltd. and OmniScriptum S.R.L publishing group

120 High Road, East Finchley, London, N2 9ED, United Kingdom
Str. Armeneasca 28/1, office 1, Chisinau MD-2012, Republic of Moldova, Europe
Printed at: see last page
ISBN: 978-620-8-27513-6

CONTENTS

DEDICATIONS

This work is dedicated to my parents, Pedro Paulo do Nascimento (in *memoriam) and* Jùlia Silva do Nascimento, for their love, dedication and unconditional support for my growth as a man and as a subject for social inclusion and professional advancement. They were the great motivators behind my growth and that of my siblings in achieving our dreams.

To my siblings who, even though far away, somehow thought of me, gave me words of encouragement or smiled with me during the happy moments of our journeys through life.

To my children Alexander de Souza Nascimento, Gleidson de Souza Nascimento, Andressa de Souza Nascimento, Geandre de Souza Nascimento, Marcos César Aguiar Sobrinho, Maria Jùlia Aguiar Nascimento, Matheus Nascimento, Juliana Nascimento, we have always shared special moments and they are the reason for my life.

Professor Luisa Helena Silva de Sousa, IFPA- Campus Itaituba, and Cicero Paulo Ferreira, tireless partners who helped directly to build this result, the fruit of a collective effort.

ACKNOWLEDGMENTS

Recognizing the achievements of those who make us grow is an art, it's a noble feeling that strengthens us and unites us around hopes and opportunities to build a brighter, more social future and so many other adjectives. So I want to say thank you:

First of all to our creator, the one who strengthens us and gives us wisdom.

To the Federal Institute of Parà - IFPA, Campus Castanhal for the opportunity given to us.

To the Federal Institute of Para - IFPA, Campus Itaituba for having supported me on this journey.

To the National Council for Scientific and Technological Development - CNPq/ MEC/SETEC for funding the research project, thus making a decisive contribution to the development of this study.

To the Coordination of the Post-Graduation Program in Rural Development and Management of Agri-Food Enterprises, in the person of Professor Dr. Cicero Paulo Ferreira, for his guidance and other teachers of the Post-Graduation Program of IFPA - Campus Castanhal, UFPA and EMBRAPA.

We would like to thank the Sao Joao Batista Community, especially José Ribamar, D. Neiva and Pedro Francisco for their unconditional support in our work program in the community that is the subject of this research.

The partner organizations, teachers, managers and students of IFPA - Campus Itaituba who supported and participated in the construction of this proposal.

PRESENTATION

The story of this book begins in a chapter that is more distant from today than the age of many readers, including this writer and the author of this book himself. In the mid-1940s, during World War II, the Brazilian government, then an ally of the United States, received the *Ford* Project to produce rubber in Brazil, with the aim of meeting the huge demand for tires for the war fleets of the time. Thus, as part of the project, the "Soldados da Borracha" (Rubber Soldiers) Program was created, promoting the arrival of Brazilian citizens from the northeast, enlisted in the armed forces, to "fight" in the Amazon jungle.

The war ends, time passes, and after 20 years, in the 1960s, during the great drought in the northeast, the military government, noting the existence of demographic voids in the Brazilian Amazon, launches the slogans "integrate so as not to deliver" and "land without men for men without land", as a way of integrating this region with the rest of the Brazilian territory, in fact, through the implementation of major roads, where one hundred kilometers on the banks of these highways would be used for the colonization of approximately 500,000 people, through the creation of family settlement agro-villages. The Transamazon highway was the main route chosen for this colonization.

Within this geopolitical context is the municipality of Itaituba, whose Transamazon Highway crosses it, making it its main land gateway, given that its main form of access is by river via the Tapajós River.

In view of the colonization resulting from the construction of the Transamazon Highway, as well as natural attributes such as a large mineral deposit and considerable reserves of forests suitable for plant exploitation, Itaituba is growing and becoming disorganized, due to the lack of adequate planning to accompany this speed of growth.

As a result of this process, the production of so-called "garbage" became an inevitable consequence, giving rise to large deposits, or dumps, designed without any environmental conditions. One such site was the Curral Redondo dump, which was the garbage disposal area for the municipality of Itaituba until the mid-2000s.

With the "decommissioning" of the Curral Redondo dump, a major problem arises: what to do with this area, which is contaminated with all sorts of products that have been deposited there for years? The answer comes from Professor Jùlio Nonato Silva Nascimento, Professor, Researcher and Coordinator of Extension and Integration at the IFPA, Itaituba Campus, who is a History major but has extensive training and experience in agriculture and regional development.

For years, Prof. Jùlio, as he is known in Itaituba, has been developing studies on the history of this region, as well as research into plants that can be acclimatized for use in the recovery of degraded areas of the Amazon. Selfless in his scientific purpose and with great sensitivity to social problems and the land, this researcher heads to Curral Redondo and begins his journey with the Sao Joao Batista Community, in the vicinity of the old garbage dump.

That's how we arrived at this incredible work to recover degraded areas, the essential principle of which is to give the community itself back its leading role in resolving environmental problems in its area.

Where the researcher guides, in addition to his students, an entire community, with the support of renowned entities in the municipality, which embrace this cause as their own.

In the pages that follow, we will take a journey through time and space, where the main focus is to give back to the Sao Joao Batista Community, an area that has been rehabilitated for the use and development of community support activities, with environmental quality and growing sustainability. In the certainty that, in a few decades' time, this area, with its new identity, will forget that it was once called a "dump".

Have a good read, and a good trip, everyone! Welcome to the community of Sao Joao Batista do Curral Redondo, in Itaituba, in the state of Parà, in the Brazilian Amazon, where rubber soldiers once fought, and where today we have lands with good men, like this researcher who is the author of this work that I am presenting to you.

Itaituba, March 14, 2017.

Liz Carmem Silva-Pereira

Profª and Researcher at IFPA, Campus Itaituba

CHAPTER I

History and Society

1.1. Building history

1.1.1. The municipality of Itaituba

In the last 40 years, the municipality of Itaituba has undergone a series of significant changes in the urban and rural areas, triggering various socio-environmental problems in view of the growing population producing solid waste that has been stored improperly, in the open, forming pockets of garbage of various kinds, directly interfering in the daily lives of the residents of the community called Sao Joao Batista do Curral Redondo, Located on the municipal road linking Itaituba to the district of Barreiras, this is indiscriminately damaging the availability of water from the small streams and springs that are established there as a natural resource, but which is unavailable for human use, according to the results of analyses by the Environmental Microbiology Laboratory of the Evandro Chagas Institute (SEMAM/IEC, 2003).

1.1.2. Everyday life and the relevant legislation

Under Law 12.305/2010, of the National Solid Waste Policy, and a tireless legal battle, the State Public Prosecutor's Office was able to get the municipality to deactivate the dump, and the municipality is obliged to recover the environment in accordance with the conduct adjustment agreement issued by the 2ª Itaituba State Public Prosecutor's Office, but this has not happened to date. As a result, the natural environment and the entire ecosystem of the area remains in a state of degradation, damaging the "passa tudo" and "Bom Jardim" streams and the springs that directly serve the people of this locality.

1.1.3. Sustainability

Conceptualizing development associated with sustainability gives us all an unconditional reflection on the environment, since some interesting questions need to be asked in order to justify our actions and needs. Thus, the question arises: How can we develop without harming? How can we justify human needs without taking advantage of natural resources? How should we relate to natural ecosystems if we need them so much? (...) Sustainable development is a concept that seems indispensable in discussions about modern societies. Since the 1960s, there have been many debates about alternative proposals regarding the risks of environmental degradation.

Recovering degraded areas is one of the main tools in this process, with a view to restoring natural resources through the use of simple and complex technologies that can help to restore the landscape by restoring the forest, air, water and soil, improving the process of water absorption in underground water tables, rainfall levels, protection of riparian forests, avoiding laminar erosion that harms streams, streams, springs and rivers, thus acting directly to improve the production chain of fauna and the balance of the climate.

1.1.4. The global world

The current circumstances manifested by the accelerated process of demands in the globalized world increasingly require technical professionals to have comprehensive qualifications that allow them to meet the demands of the competitive market, valuing the dynamics of everyday life, taking into account the challenges caused by the development of productive activity, as well as those of sustainability in relation to the multiplicity of productive processes organized through the growing Cooperative association and the relationship between man and nature.

(GAIGER apud Laville, 1994) comments that "the appearance, on a growing scale, of popular enterprises in free association, cooperative work and self-management, is today an indisputable fact in our social landscape, as well as being a phenomenon observed in many countries".

According to Jesus (2007), in Brazil, from the second half of the 19th century onwards, any form of organization along the most diverse lines of interest had to apply for permission from the police station and he constructs a historical conception of associative forms and currently, this associative phenomenon in Brazil has spread with relative intensity.

1.1.5. Community organization and the impacted ecosystem

The Sao Joao Batista community, the area that is the subject of this research, is located 11 kilometers from Itaituba and is part of the municipal road that connects Itaituba to the district of Barreiras. It concentrates a great potential for agricultural exploitation, involving small and large properties that bring together associations of organized family producers and large beef cattle ranches in an extensive breeding system.

The research was carried out *on the basis of on-site*, investigative activities, with the aim of answering the proposed questions, in order to minimize the effects of the aggression to the natural environment caused by the presence of solid waste dumped by the municipal government in the Sao Joao Batista community of "curral Redondo", which over the last two decades has received urban waste in an area of approximately 22 hectares. The extent of the damage caused to the springs that feed the micro-basin of the "Passa Tudo" stream is clear to see, compromising and polluting the surface waters of this stream with leachate (waste resulting from organic decomposition).

The pre-existing environmental conditions reveal a lack of effective public policies that can minimize the worsening problem, compromising the health of the local population and the entire region that is directly influenced by this water corridor. For this reason, we have tried to articulate and propose to the community, using an alternative system, the recovery of the area, with the starting point being the afforestation of the impacted area, with the planting of previously studied forest essences, native fruit trees (wild plants) and the use of creeping leguminous plants.

As such, we take as our starting point the current understanding of degraded areas, which has as its conception those that present "symptoms" such as "erosive processes, absence or reduction of vegetation cover, deposition of garbage, soil compaction, silting up of rivers, among others" (CURY & CARVALHO 2011).

The aim of this research was to implement an alternative system for the recovery of degraded areas with an emphasis on the recovery of springs, using seedlings of forest essences, fruit trees and seeds of leguminous plants in the anthropized area of the former solid waste dump (lixao), in the Sao Joao Batista community, in the municipality of Itaituba - Parâ.

Given these precepts, it was possible to build the instruments that would enable the study to be carried out and the logical sequence of the work to be realized, as listed below:

- A participatory diagnosis of the community was generated involving the socio-economic, biophysical and land aspects of the community;
- The implementation of an alternative environmental education system was made possible, enabling changes in the community's daily practices;
- A nursery was set up to produce seedlings;
- An experiment was carried out with seedlings of Açai comum (Euterpe oleracea m.), Paricà (Schizolombium amazonicum) and Fava arara (Parkia multijuga), to analyze the results in relation to the amount of macronutrients (N. P. K.) absorbed by them over a period of up to 4 months;

- A forest recovery system was implemented with the planting of forest species: Paricà (Schizolombium amazonicum), Jatobà (Hymenea courbaril L.), Andiroba (Carapa Guianesis aubi), Brazilian Mahogany (Swietenia macrifylla king). Native fruit trees: Buriti (Mauritia flexuosa) and Açai (Euterpe oleracea m.). And leguminous plants: Pork bean (Canavalia ensiformis); Guandu bean (Cajanus cajan); Black mucuna (Mucuna pruriens). With a view to minimizing the impact of human action and nature on the environment;
- Alternative technologies were developed to improve the environment and restore the springs by planting riparian vegetation and building containment barriers.

CHAPTER II

The paths of history, economic and social changes

2.1. Historical and social changes since the industrial revolution

With the advent of the Industrial Revolution, men have historically focused their actions on the urban environment, producing countless artifacts that result in the production of solid waste that transforms the natural landscape and pollutes the environment. This panorama, according to Neto (2013), "allows us to identify the profound correlation between waste generation, the way of life adopted and the economic activities carried out in each community". Brazil is no different. According to Guerra & Cunha (2013), "urban concentration in Brazil is around 80% of the population and its development has been carried out in a poorly planned manner, with major institutional and technological conflicts". The authors also argue that the main problems are related to major floods and the impact on the environment and that due to a lack of planning, cities are extremely compromised, generating chaos and a high social cost. For Luckesi & Passos (2012):

> "In everyday life, knowledge seems to be something so commonplace that we don't ask ourselves what it is, its process, its origin, its form of appropriation. Little by little, throughout our childhood, adolescence and youth, we acquire an understanding of the things that make up the world around us, of relationships with people, of the moral and social norms that govern relationships between human beings."

In this way, we need to build an understanding that provides us with feasible means of minimizing environmental impacts in the 21st century, which imposes on us an arduous reflective mission, leading us to rethink our actions on the environment in which we are inserted and, to this end, we need more consistent and committed public policy proposals that allow us to emerge to new horizons; It is also necessary to develop scientific research, given the complexity of environmental impact processes, as Coelho (2013) points out, emphasizing that urban environmental impact presents a double challenge. "On the one hand, it is necessary to problematize reality and construct an object of investigation". "On the other hand, it is necessary to articulate a coherent interpretation of the ecological (biophysical-chemical) and social processes involved in the degradation of the urban environment." He argues that the challenge of problematizing and constructing an object of research is challenged by a process that transforms the problem of urban environmental impacts. This leads us to rethink our practices in the world of science and its fields of research. For Derisio (2012):

"Every living being reacts with its environment and produces waste. Unless the environment can dispose of them properly (self-depuration), these wastes can interfere with the life cycle. Living in communities, human beings have developed processes that produce large quantities of by-products or waste in the form of material or energy. This process has profound economic, political, social and health significance, affecting the health of man himself, his comfort and security, his wealth and his power".

Based on this assumption, environmental pollution is human action, since everything that is thrown into the environment or released into the water, air or soil in any form of matter or energy, frequently and in concentrated volumes or in situations that differ from what is considered to be the quality standard established by law, becomes a pollutant of the environment, such as pollutants in their various forms: physical, chemical, physical-chemical, biochemical, biological and radioactive, directly affect natural ecosystems and cause changes in the environment. He goes on to emphasize that "the various forms of pollution are interconnected, so that pollution control must be done together", since a polluting action on the soil can influence the quality of surface water or groundwater if certain actions are not taken.

2.2. Sanitation in Brazil

In Brazil, the configuration of the basic sanitation sector began in the 1970s, with the establishment of the National Basic Sanitation Plan (PLANASA) by the federal government (NETO, 2013, p.14), mentions that the institution's experience was successful in generating a rapid expansion of sanitation services in Brazil, but insufficient to meet the demands generated by the country's growing urbanization process, which without a National Basic Sanitation Policy finds it difficult to regulate and universalize services. He adds that even with the creation of the Ministry of Cities in 2003 and the perception of integration between urban policies, with emphasis on the National Secretariat for Environmental Sanitation, policies were still disjointed and after several attempts to create national legislation on basic sanitation, Law No. 11.445/2007 was approved, establishing national guidelines for the issue.

"This legal framework also stands out for encouraging the associated management of basic sanitation services, developed through cooperation agreements or public consortia. However, despite establishing as one of its guidelines the stimulation of the implementation of infrastructures and services common to municipalities, through cooperation mechanisms" (BRASIL, 2007[a] , art. 48).

According to Neto (2013), the management of urban waste is essentially a municipal activity and the activities that make it up are restricted to the territory of the municipality, so combined solutions are not very common in Brazil, even when it comes to final disposal in landfills. He concludes, "Urban solid waste management in Brazil reveals the seriousness of the current situation, calling for an improvement in the efficiency of the services provided in order to reduce the impact on the urban ecosystem and preserve the public health of the population."

Basic sanitation is the part of the environment that focuses specifically on the following services: water supply; sanitary sewage disposal; waste conditioning, collection, transportation and disposal. These are extremely important services for a given community, given that each person, on average, produces 1 kg of household waste every day and uses, on average, 150 to 200 liters of water for food and hygiene (figures for around 50% of the population that has access to water supply services); each person produces, in terms of sewage, the equivalent of their daily water consumption (CARVALHO & OLIVEIRA, 2010, p.20).

Magossi and Bonacella (2013) conceptualize the importance of water in the phrase "water is one of the most important elements for all forms of life on Earth". "Without it, no one would survive and it is present in all living organisms along with a multitude of substances." The authors also infer that in nature water can be found in three different physical states: solid state (in the form of ice) making up approximately 2% of the total volume of water on the planet. In the liquid state, it represents no less than 98% of the entire volume of water on Earth. This is the state in which living organisms use the most, the vast majority (97%) of which is made up of water from the seas and oceans and just over 1% refers to continental water, which can be surface (rivers, lakes and lagoons) or underground, deposits of water called water tables. The gaseous state, in the form of vapor, in the atmosphere and in some underground deposits in volcanic regions, represents only 0.005%, but plays a fundamental role, especially for the atmosphere. According to Carvalho & Oliveira (2010):

> "As you can see, water - a liquid *already* called the "blood of life" and just as precious - is actually a finite and vulnerable resource. However, man does not seem to have realized this yet, as the environmental degradation caused by the pollution of water resources ends up contributing to the worsening of the problem of water scarcity, represented by the difficulties in supplying large cities and industrial areas, and the supply of water for irrigation, an obstacle that can occasionally be associated with periods of drought."

According to Santos et al. (2009) "the multiplicity that makes up the environment, together with the diversity of interrelationships, is enough to infer a judgment on the complexity of the environmental dimension that has emerged as a crisis of civilization".

In this context of problematization, the fundamental and transforming element for the authors is Environmental Education and the Paradigm Shift, because in the 21st century, little has been done to resignify this shift, since several factors contribute to the environmental problem correlated with other crises in third world countries such as the economic, arms and energy crisis, which degrades the lives of these subjects and the problem in the big cities of the most industrialized countries. To this end, the paradigm shift, which calls for new ways of breaking away from sociological determinisms, calls for education, as a social phenomenon, to evolve in this direction. "By modifying the environment to enable his most diverse activities, man has systematically caused impacts whose effects are not always immediately visible" (ROSSETO & LERIPIO, 2012).

CHAPTER III

Today's society and its evolution

3.1. Implications of Antropic Action

For Araújo (2012), "three and a half billion years of evolution have resulted in the current wealth of life on our planet, traditionally measured by the number of species of living organisms". He goes on to emphasize that this does not adequately represent the extraordinary variety and complexity of nature, which leads us to reflect on biodiversity at different levels that are established by genetic diversity, diversity of species, diversity of communities and ecosystems that interact with the physical environment through processes of energy and element exchange, such as photosynthesis, respiration and biogeochemical cycles (IBIDEN). It also states that some authors consider human cultural diversity to be a component of biodiversity, manifested in the variety of languages and dialects, religious beliefs and land management practices, art, music, cuisine, social structure and so many other attributes of human society.

For the aforementioned author, since Brazil is predominantly a tropical country, its biodiversity has a direct influence on the planet since this area lies between the Tropics of Cancer and Capricorn, which cover around 40% of the planet's surface. From an ecological perspective, the tropics comprise the region delimited by the 20°C average annual temperature isotherm, which poses enormous challenges for the conservation and management of tropical ecosystems (ARAÙJO, 2012).

Still in this ecological direction, according to Carvalho & Oliveira (2010), "Environment is characterized by the space in which the interactions of living beings with each other and with the environment in which they live occur". (...) in this conception, the analogy made by the author is pertinent, when they emphasize the profound imbalances and growing environmental degradation caused by human intervention in recent decades, the transformations have been such that they have led man to understand that the world is one and every aggression that causes imbalance and devastation anywhere on the planet can compromise the environment as a whole. He reiterates that global problems, such as climate change and the destruction of the ozone layer, end up directly or indirectly affecting various parts of the earth.

For Philippi Jr. & et al, (2012), "the development models adopted historically in Brazil have resulted in social, economic and environmental impacts that have caused an excessive concentration of income and wealth", which means in practice the great regional differences and the significant exclusion of a large part of the population from the sharing of these riches, concentrating opportunities in the hands of a few and marginalizing the majority who are subjugated to capitalist politics, without options or alternatives that could be fairer and more equal in the process of distributing income for citizen life.

Ecology as a science studies the environment, Ecology as a science studies the environment, seeking to understand the multiple and complex interactions that the various species of living beings establish between themselves and with the environment in which they live. It is therefore necessary to take care of the balance of nature, which presupposes knowing and respecting the ecosystem and only through an educational process will it be possible to find ways to improve the quality of life of the population through actions that guarantee development with principles of sustainability and the conscious formation of citizens capable of exercising social citizenship and promoting continuous transformative actions, with the aim of reversing the existing process of socio-environmental degradation (CARVALHO & OLIVEIRA, 2010).

"Brazil is considered to have the greatest biodiversity on the planet. Given its continental size and great geomorphological and climatic variation, it is home to six terrestrial biomes: the Amazon, the Cerrado, the Pantanal, the Atlantic Forest, the Caatinga, the southern grasslands and three large marine ecosystems, which include eight marine ecoregions" (BRASIL, 2010).

Conceptualizing a degraded area requires prior knowledge that is directly related to the means of regeneration that exist in the environment, especially vegetation and its biotic means, such as seed banks, seedling banks, seed rain and regrowth, which we are constantly witnessing in the various regions of the planet through human intervention with the accelerated and disorderly growth of cities and the increase in human and animal population.

"Brazil, in its current phase of socio-economic development, must face the challenge of educating its population and training highly qualified human resources to consolidate a science and technology policy committed to the permanent construction of scientific, technological and innovation bases required for social, environmental, economic, spatial, political and environmental sustainability" (CAPES, 2012).

With the growth of the population and the disordered use of natural resources, the situation of the planet is becoming increasingly serious, imposing intense pressures on us to address "energy, metropolitan areas, environmental protection and recovery, expansion and improvement in food production, monitoring and forecasting of external meteorological phenomena and vulnerability, use and protection of water resources, exploitation and conservation of biodiversity, as well as the development of products and services" (CAPES, 2012).

In the history of mankind, according to Araùjo (2012), a large part of human progress has derived from the exploitation of biological resources involving food, pharmaceutical and medicinal products and natural fishing resources.

Biological diversity is of great importance to Brazil. According to IBGE data (2010), the national harvest was a record 149.5 million tons of grain, representing 7.1% of world production (IBGE, 2010). (...) biodiversity is an important component of ecological systems, from which we derive a series of goods and services that contribute decisively to the well-being of humanity.

Forest fragmentation reduces the total area covered by a forest, which can result in the extinction of some species. In this way, studying landforms is a fundamental aspect of problematizing degraded environments in terms of how they can be maintained and recovered. In this sense, human action will determine the various activities on the surface of the soil, which will justify the greater or lesser intensity of the response, which may or may not reach significant proportions (catastrophe), which will depend directly on the use and management of the soil and its physical characteristics (GUERRA, 1999 and 2001; GOUDIE, 1995).

"Reliefs are the floors on which human populations settle and carry out their activities, deriving economic and social values from them. Depending on their characteristics and the processes that act on them, they offer people the most varied types and levels of benefits or risks. Their greater or lesser stability also stems from their evolutionary tendencies and the interference they may suffer from other environmental components or from human action" (SILVA apud MARQUES, 2001, p. 13-18).

In this relationship, nature always has answers that permeate the intensity of anthropogenic action, which historically has shown us a series of catastrophes, the result of human recklessness. Therefore, the study of relief enables us to prevent such occurrences and helps us to find the best way of recovering a degraded environment.

"The specialist working on the recovery of degraded areas must recognize the different materials on which the work will be carried out, identifying them as being more or less resistant to the action of external agents. For example, soils with a higher content of fine sand and silt are more susceptible to erosion than clayey soils, considering only the texture" (LUK, 1979).

According to Carvalho & Oliveira (2010), "soil is the top layer of the earth's crust formed naturally from chemical, physical and biological processes, such as the decomposition of organic matter, the fragmentation and decomposition of rocks, the dissolution of chemical substances, water and air". (...) the climate, rainfall, temperature and humidity, topography, the diversity of fauna and flora, the origin and age of the rocks, are all factors that modify the composition of the soil.

Another significant factor that has influenced the transformation of the soil is human action through the different types of exploitation and use of the soil, whether through productive activity or the use of extracted natural resources. Its average proportion of components is as follows: 45% mineral elements, 25% water, 25% air and 5% organic matter and the most frequent alterations are established as follows: a) changes in soil structure due to agricultural and livestock activities; b) soil sealing due to the construction of paved roads, buildings in general, sidewalks and others; c) earth movement due to excavations and the construction of landfills; soil degradation due to the improper disposal of solid and liquid waste, the use of pesticides (insecticides, herbicides, fungicides, etc.), the use of fertilizers, etc.), the use of synthetic fertilizers, as well as burning and erosion; d) salinization.

The principle of soil degradation is directly related to the various actions of man that modify and transform nature and also through natural phenomena causing environmental impacts that alter the natural landscape (CARVALHO & OLIVEIRA, 2010), (...) as a result of these actions, we will find deteriorated soils, in some cases unrecoverable from a productive point of view.

In practice, the erosive action of the most superficial particles of the earth's surface is called laminar erosion. The second stage is called furrow erosion and the aggravating phase is called gullies, factors that occur in the natural environment or as a result of man's intervention, when he uses the natural resource improperly and irresponsibly. For example, deforestation has been the great villain of the process, as it contributes alarmingly to erosion. In bare soil, it becomes easy for water to emerge as the next factor in the devastation and change of the landscape caused by erosion (IBIDEN).

"The lack of vegetation cover is one of the main causes of erosion. Creeping plants, trees, shrubs and all kinds of vegetation are the soil's greatest protection against erosion. Vegetation reduces the impact of raindrops on the soil, minimizes the effect of particles being carried away and reduces the speed of water runoff" (CARVALHO & OLIVEIRA, 2010).

3.2. Disorderly Occupation of Urban Areas

According to Braga et al. apud Miller (1985), "our planet can be compared to an astronaut", traveling at a hundred thousand kilometers an hour through outer space, without the possibility of stopping to refuel, but with an efficient system for using solar energy and recycling matter". What the author reveals to us by exemplifying this astronaut is that the planet has a certain amount of water and food for a certain population and as this number of passengers grows, in a way, water and food will become more difficult and that if we don't reflect on this issue, disastrous consequences will arise, causing total imbalance.

The authors refer to the growth of the world population from 2.5 billion people in 1950 to 6.2 billion in 2002 (U.S. Census Bureau, 2004) and currently, the growth rate is around 1.13% per year. Making an analogy with an airplane, it carries 6.2 billion passengers and every year, another 74 million passengers board it and, considering the population problems that occur in less developed countries, this imbalance tends to get even worse.

When we analyze the figures, we realize that they are alarming, because within this perspective of growth, we must ask ourselves how long natural resources will be sufficient to sustain all this population? This is why institutionalized and strengthened public management is necessary, as Vidal (2012) points out: "the role of institutions in driving change has been interpreted from different perspectives", leading some authors, from an interactionist reading, based on the conscious action of men, to state that they are products and producers of institutions.

According to Mota (2010), population growth in the last two decades has been considerable and he takes a numerical approach by referring to the years 1850 to 1930, showing that in these eighty years the population doubled while in the years 1970 to 2010, that is, in just 40 years there was a doubling of the world population. This means that human intervention in the environment is only set to increase and, therefore, there is a need to change the paradigm of development at any cost.

The above-mentioned author discusses the growth of the Earth's population from various points of view, demonstrating the greater concentration of people in developing countries (4.2 billion in 1990 and 4.9 billion in 2000), as opposed to developed countries (1.2 billion in 1990 and 1.2 billion in 2000). As a result, more people are living in precarious conditions of housing, food, education, employment, health and sanitation. Problems with the big cities, which are receiving more and more inhabitants, and basic infrastructure conditions are not keeping pace, generating serious social and environmental problems.

The growth of wealth-generating activities leads to an increase in the use of natural resources (water, air, soil, vegetation, fauna, energy), causing sudden and intense changes in the environment. "The production of waste, in solid, liquid, gaseous or energy forms, which is released into the lithosphere, hydrosphere and atmosphere, often in large quantities, does not allow ecosystems to recover, damaging their components, including man" (MOTA, 2010).

"In his evolutionary process, man has faced various situations that have taught him to observe nature in order to use it to his advantage and provide for himself. Thus, mankind has moved through different patterns of social organization and production systems, with health actions being a valuable tool in characterizing the political structure in force at each period, and vice versa" (REZENDE & HELLER, 2008, p. 52),

3.3. Solid Waste and Final Disposal

According to Oliveira (2009), in order to discuss and analyze the problem of solid waste in large urban centers, it is necessary to take a historical approach to waste production. His perception of the problem is marked by the advent of the industrial revolution in the world, especially in Brazil, where from the 1950s onwards, with the use of packaging, waste production increased considerably. According to data from UN-HABITAT (2009), Neto (2013) is emphatic in saying that the world produces five million tons of solid waste a day, based on the growing consumption of non-durable goods and the growing population index in cities. He also points out that cities are built on ecosystems that have a limitation and are therefore correlated between the resources used and the capacity of ecosystems to supply them without altering their natural balance.

(...) "The management of solid urban waste has emerged as an issue of fundamental importance, demanding an efficient management system from public bodies, involving the treatment of discarded materials, as well as the environmentally appropriate disposal of waste, reducing impacts on urban ecosystems" (Neto, 2013, p. 1).

In Brazil, according to (Neto apud Phillipi Jr; Aguiar, 2005) the government's concern only began in the 20th century, between the 1940s and 1970s, influenced by high rates of demographic growth and migration to the cities, reaching exorbitant numbers, in the order of 34 million inhabitants in urban areas, while the necessary public services such as infrastructure did not keep pace with this expansion.

Although the Federal Constitution of 1967 made it possible to give responsibility to the states (metropolitan regions), this was only implemented during the military regime, and was characterized by an imposing bias aligned with a centralizing policy. This fact lasted until 1988 when, after the country's re-democratization process, a new Magna Carta was instituted for the country.

The author cited above, who also refers to (Alves 2007; Moura et al. 2004; Klink 2009), points out that until then the systems in place have remained obstacles to the consolidation of models capable of responding to the shared management of the territory and this is reflected to this day because even with the National Solid Waste Policy (Law 12.305/2010), many Brazilian municipalities still do not store their waste in appropriate environments called landfills. In this way, they continually pollute the environment. "(...) solid waste in Brazilian legislation with Agenda 21 (UNCED, 2001), with NBR 10.004/2004 (ABNT, 2004) and with US legislation (USA, 1989 apud Phillip Jr.; Aguiar, 2005), there is a clear conceptual alignment, where all start from the concept of waste as a by-product of anthropic activity with specific characteristics inherent to its constitution process" (Neto, 2013).

The same author describes that solid waste is classified according to the process that originated it as well as its constituents and characteristics; he establishes categories ordered according to their hazardousness: Hazardous waste (Class I), Non-inert waste (Class II-A) and Inert waste (Class II-B). In addition to hazardousness, several authors have listed multiple categorization criteria that can be used, including classification by origin (Phillip Jr.; Aguiar, 2005; Jardim, 2000; Lima, 2001; Monteiro & et al. Castilhos Jr. 2003; Brasil, 2010).

For Carvalho & Oliveira (2010), Solid Waste according to NBR 10.004/04 is defined as: "Solid and semi-solid waste resulting from industrial, domestic, hospital, commercial, agricultural, service and sweeping activities". He goes on to emphasize that "solid waste contributes directly and indirectly to environmental pollution" and that this whole chain of contamination of soil, water, animals, food and those who handle the waste favors the appearance of diseases such as diarrhea, leptospirosis, verminosis, cholera, typhoid fever, salmonellosis, trichinosis, cysticercosis and others. He adds that "when dumped in the sea and rivers, waste contributes to polluting them, destroying the environment and marine and freshwater food reserves".

With a view to consolidating public policies for the environment, the Brazilian Forest Code (Law 4.771/1965), in its capital VI, Art. 225, reads as follows: "Everyone has the right to an ecologically balanced environment, which is a good for the common use of the people and essential to a healthy quality of life, and the public authorities and the community have the duty to defend and preserve it for present and future generations" (BRASIL, 2010, p.14).) and presents specifics for this guarantee such as: in paragraph 1, item VI "promote environmental education at all levels of education and public awareness for the preservation of the environment"; VII, "protect fauna and flora, prohibiting, in the form of the law, practices that jeopardize their ecological function, cause the extinction of species or subject animals to cruelty" and paragraph 4, "The Brazilian Amazon Forest, the Atlantic Forest, the Serra do Mar, the Mato Grosso Pantanal and the Coastal Zone are national heritage, and their use will be made, in the form of the law, within conditions that ensure the preservation of the environment, including the use of natural resources".

Article 2 of Law No. 6.938 of August 31, 1981, on the National Environmental Policy, reads as follows: "The purpose of the National Environmental Policy is to preserve, improve and recover the environmental quality that is conducive to life, with a view to ensuring, in the country, conditions for socio-economic development, the interests of national security and the protection of the dignity of human life" and lists the following principles:

I - Government action to maintain ecological balance, considering the environment as a public heritage to be necessarily ensured and protected, with a view to collective use;

II - Rationalization of the use of soil, subsoil, water and air;

III - Planning and monitoring environmental resources;

IV - Ecosystem protection, with the preservation of representative areas;

V - Control and zoning of potentially or effectively polluting activities;

VI - Incentives for the study and research of technologies aimed at the rational use and protection of environmental resources;

VII - Monitoring the state of environmental quality;

VIII - Recovering degraded areas;

IX - Protection of areas threatened with degradation;

X - Environmental education at all levels of education, including community education, with the aim of enabling the community to actively participate in the defense of the environment.

"The environment is the space in which living beings interact with each other and with the environment in which they live. Not so long ago, the environment tended to be seen only as the neighborhood, the surroundings. However, the profound imbalances and growing environmental degradation caused by human intervention in recent decades have led people to understand that the world is one and that the imbalance and devastation occurring in certain parts of the planet can compromise the environment as a whole" (CARVALHO & OLIVEIRA, 2010, p. 10).

The origin of solid waste is one of the key factors in understanding the degree of environmental damage. In this way we can characterize them as: urban, industrial, health service, airport and port, agricultural, radioactive, road and rail transport loading and unloading areas and rubble (CARVALHO & OLIVEIRA, 2010). The authors refer to the constant production of this waste, inferring the problem of production by emphasizing that urban waste does not represent the largest amount of waste, but rather that which comes from industry in the most diverse forms: "solid, semi-solid, gaseous and liquid, with final disposal, if possible, in the public sewage system or in bodies of water, according to CONAMA (National Environmental Council) Resolution No. 313, of 2002". (...) "Industrial waste has a specific classification according to its peculiarity (NBR 10.004/04) and is divided into two classes, according to procedures and tests established by ABNT standards (NBR 10.005/04 - Waste leaching / NBR 10.006/04 - Waste solubilization / NBR/10.007/04 Waste sampling" (ibid).

CONAMA Resolution No. 258/99, in its Article 9, "states that as of its entry into force, the inadequate final disposal of unsalvageable tires is prohibited, such as disposal in landfills, the sea, rivers, lakes or streams, vacant lots or swamps, and open burning" (ARAÙJO, 2008, p. 111).

According to Neto (2013), studies show that large cities and developed countries are the biggest producers of solid household waste, since the consumption that generates disposable packaging and the rapid obsolescence of objects and equipment leads to an ever-increasing level of production and consumption.

This table shows an overview of the correlation between solid waste production, the way of life adopted and the economic activities carried out in the 14 largest Brazilian cities Sao Paulo, Porto Alegre, Belém, Salvador, Belo Horizonte, etc. (0.2% of the total) were responsible for producing 19.06% of the total volume of solid urban waste generated in the country in 2008 (IBGE, 2010).

In Brazil, according to Neto (2013); Carvalho & Oliveira (2010), solid waste production is estimated at around 240,000 to 259,500 tons per day. This view, according to the authors, is worrying, since "of the total waste collected per day in Brazil, around 35% was still disposed of in open dumps or controlled landfills in 2008", which could be minimized if the final destination complied with the basic precepts of selective collection and recycling, The treatment of waste currently includes both solutions applied at the source of waste production and those implemented by public collection systems" (CARVALHO & OLIVEIRA, 2010) and can be characterized as follows: a) treatment at source, building shredding and building incineration; b) treatment in the public system, recycling, sanitary landfills and energy landfills, incineration, aerobic digestion in composting plants and dumps, which are incorrect alternatives because they contaminate the environment with the production of leachate, resulting from the decomposition of waste. "The growth of the urban population, accompanied by industrial development and other waste-producing activities, increasingly demands that actions to control environmental pollution be improved and intensified" (MOTA, 2010).

CHAPTER IV

Technologies for Recovering a Degraded Environment

1.1. Recovering Springs with Vegetation

For Mota (2010) "In the biosphere there are large ecosystems, whose characteristics depend on their geographical position, their geological history and the biological evolution of the planet". He emphasizes that each ecosystem has its own essential elements, such as inorganic and organic substances, climate, autotrophic and heterotrophic organisms. These large ecosystems can also be called terrestrial or aquatic biomes.

"Forests are large areas covered with dense vegetation, composed mainly of large trees and shrubs. A variety of vegetation types can be found in a forest, from herbaceous plants to trees, adapting to various soil and humidity conditions" (MOTA, 2010).

According to Cury & Carvalho Jr. (2011) "the concern with repairing damage caused by man to ecosystems is not recent and is currently emerging as an important challenge to be overcome". He adds that forest plantations have been carried out in Brazil for various reasons since the 19th century. However, only since the 1980s, with the development of the science of restoration ecology, have they been more clearly defined.

According to Guerra & Cunha (2013) "the process of scientific research requires a close relationship between theory and research methods". And they stress that great care must be taken with the methods of investigation and interpretation, as well as with the relationships between the theoretical and the real object, whether or not they are operationalized by research methods and techniques.

In order to organize an understanding of the process of restoring springs, it is necessary to take a preliminary look at the importance of water in the context of life in large natural reservoirs (oceans, lakes and rivers), since there are various living organisms there, from large fish and mammals to tiny single-celled beings, the protozoa. Among the plants, large algae and even microscopic algae play a vitally important role for life on earth (MAGOSSI & BONACELLA, 2013).

Water is a fundamental component in the dynamics of nature and plays a key role in almost all life cycles and human activities (TREVISAN, 2010). In Brazil, according to Magossi & Bonacella (2013), despite having around 12% of all surface fresh water on the planet, many people still suffer from the scarcity of this precious commodity and, according to data from the Ministry of Cities' National Sanitation Information System (SNIS, 2010), around 19% of the Brazilian population still does not have a piped water system.

Another important piece of information is provided by the Ministry of the Environment, which presents data on the hydrographic basin in the south and southeast of the country, showing that 70% of these reservoirs are contaminated by industrial or domestic waste.

Another interesting aspect is the distribution of this water across Brazil. The North concentrates around 70% of the country's surface freshwater, the Southeast concentrates 43% of the total population and only 6% of water resources; the Northeast, 29% of the population and only 3% of water (MAGOSSI & BONACELLA, 2013). When we come to understand this reality, we understand that there must be a greater commitment on the part of public policies to the cause and organized civil society to strengthen these relationships in everyday life. In this way, the recovery of springs is essential, because they play an extremely important role in balancing life in ecosystems.

The theoretical basis for implementing forestry systems will be consolidated technologies, analyzing the relationship between water and soil. This interaction, according to Trevison (2010), "governs large parts of the processes of contamination and purification of water, the recharging of aquifers and, consequently, of springs, rivers and lakes, as well as the phenomenon of water erosion". (Pregelli et al. apud Rodrigues & Gandolfi, 2000), corroborates the recovery of degraded areas:

"As a strategy or model for conserving and maintaining biodiversity, abandoning the old concepts in which restoration techniques were based on agricultural or silvicultural practices of planting perennial tree species, towards a more systemic vision, with restoration involving numerous areas of knowledge seeking to restore the environment to its resilience" (Pregelli et al. apud Rodrigues & Gandolfi, 2000, p. 1-6).

Therefore, the aspects of agroecology systematize this understanding, when it refers to the need for interaction and articulation of knowledge and know-how between the various sciences which, in an articulated way of knowledge and know-how, play a vital role in understanding and proposing ways to minimize the phenomena that surround the environment (CAPORAL et al.2006), (...) uncovered areas reduce their capacity to retain rainwater, due to constant leaching, which will not allow the water table to be adequately supplied, reducing the amount of water stored and thus interfering with the most critical periods in the springs. Areas that have been anthropized with solid waste will be sources of contamination of the water table, which will consequently generate more pollution (CURY & CARVALHO JR,

2011).

Law No. 7.754/1989, in its Article 1, reads as follows: "Forests and other forms of natural vegetation existing at river sources are considered permanent preservation under Law No. 4.771, of September 15, 1965". Law No. 8.171/1991, Art. 1, IV, "to protect the environment, ensure its rational use and encourage the recovery of natural resources" and Art. 4, IV, "environmental protection, conservation and recovery of natural resources" and Art. 20, V, "river basins are the basic units for planning the use, conservation and recovery of natural resources" (BRASIL, 2010).

For Silva (2014), environmental restoration can be understood as the "restitution" of a degraded ecosystem or wild population to a non-degraded condition, which may be different from its original condition. (...) "Restoration is the act of making a degraded area useful, usable again, but for this it is necessary to have detailed planning of the entire area from its past and present history". For Cury & Carvalho, (2011) "forest recovery projects can be planned in various ways, but they should always start with an assessment of the situation of the degraded area".

"Recovering degraded environments or ecosystems involves measures to improve the physical environment, for example: the condition of the soil, so that vegetation can be established, or the quality of the water, so that biotic communities can be re-established, management measures, the biotic elements of the ecosystem such as planting seedlings of tree species or reintroducing fauna" (SILVA, 2014).

Environmental restoration is generally carried out through the application of management technologies, which are based on making the degraded environment suitable for a new productive use, provided it is sustainable. (...) they can lead to a better environmental condition than the initial situation (a previously altered environment), (SILVA, 2014).

"Nature conservation: the management of human use of nature, comprising the preservation, maintenance, sustainable use, restoration and recovery of the natural environment, so that it can produce the greatest benefit, on a sustainable basis, for current generations, while maintaining its potential to satisfy the needs and aspirations of future generations, and ensuring the survival of living beings in general" (BRASIL, 2010).

Art. 2 of Decree No. 3.420/2000, which creates the National Forest Program, sets out the essential objectives of the program and lists them as follows: I - to encourage the sustainable use of native and planted forests; II - to encourage reforestation activities, especially on small rural properties; III - to recover permanent preservation forests, legal reserves and altered areas (BRASIL, 2010).

This understanding leads us to understand the importance of vegetation in the ecosystem, whether wild or planted, which plays special roles in the food chain of species, directly and indirectly contributing to balance and sustainability (MOTA, 2010).

According to Lorenzi (2009), wild species have been interacting with the environment for thousands of years and "have undergone a rigorous process of natural selection that has generated genetically resistant species adapted to our environment". He points out that "exotic species" brought from other countries have not undergone this process and under no circumstances are they an ideal substitute for native vegetation in all the functions it performs in the ecosystem.

"Native forests absorb excess rainwater that eventually runs off the surface of the soil, preventing it from reaching streams and rivers. In this way, the water is returned to the atmosphere by the transpiration of the trees, forming new rains. The absence of forest would allow this water to be lost to the rivers and finally to the ocean, where it would be difficult for it to return in the form of clouds to form rainfall. Therefore, the existence of islands of native forest, especially on steep slopes and along rivers and streams (riparian forest), is fundamental to keeping water resident throughout the interior of the territory" (LORENZI, 2009, p. 5).

Forests, especially riparian forests, play an important role in the dynamics of the ecosystem, protecting the soil from erosion, siltation, compaction and sealing, especially those caused by the urbanization process, which generate a series of problems for the environment (MOTA, 2010) and play vital roles in the quality of water in springs: They absorb and filter rainwater, prevent contamination of springs and increase the supply of unpolluted water to underground aquifers (LORENZI, 2009). He adds: "Wild forest, especially on steep slopes, hilltops and along rivers, streams and dams, acts as an obstacle to the free flow of water from downpours, reducing its speed and allowing it to infiltrate the soil for absorption by plants and to feed underground aquifers". Wild plants make a decisive contribution to preventing the silting up of the beds of streams, rivers, estuaries, lagoons and floodplains (ibid).

"Native forests shelter and feed fauna and ensure their diversity. In this way, they encourage an increase in the number of natural enemies of pests on surrounding agricultural crops. They also provide shelter for pollinating agents, which play an important role in improving the quality and quantity of agricultural products" (LORENZI, 2009).

Begon (2007) corroborates this by systematizing the importance of atmospheric nitrogen fixation in mutualistic plants and states that "the nitrogen-fixing bacteria found in symbiosis (not necessary mutualistic) are members of the following taxa" Rhizobia, Actinomycetes of the genus Frankia, Azotobacteraceae, Bacillaceae, such as Clostridium spp, Enterobacteriaceae, such as Enterobacter and Citrobacter, Spirillaceae, such as

Spirilumlipiferum, Cyanobacteria of the Nostocaceae family, which fix nitrogen in the root nodules of most legumes and non-legumes and thus contribute to improving the nitrogen level in the soil.

"The inability of most plants and animals to fix atmospheric nitrogen is one of the greatest enigmas in the process of evolution, as this element has a limited supply in many habitats. However, the ability to fix nitrogen is widespread, although unevenly distributed, among eubacteria ("true" bacteria) and archaeobacteria, and many of them have been observed in stable mutualisms with systematically quite different groups of eukaryotes. It is assumed that such symbioses evolved in independent epochs. They are of enormous ecological significance because of the frequent importance of nitrogen" BEGON, 2007, apud SPRENT and SPRENT, 1990, p. 401-404).

1.2. Environmental Education

Environmental education sets its primary objectives for the quality of life of the population, with special purposes, with a view to guaranteeing sustainable development, forming citizens who are able to exercise their social citizenship, that is, so that they can develop continuous transformative actions, with the aim of reversing the disturbed picture due to human action on the environment (CARVALHO & OLIVEIRA, 2010).

For Arrazola & Thiollent (2014), the problem of education in its guidelines for action with the population is characterized as an educational process, as education. This shows that the term is related to a social process and is marked by a contradiction, in a society like ours divided by social interests in struggle.

"The process of knowledge, as a dynamic reflection of objective reality, is also real (material), just as much as any corporeal substance, property of matter or other form of materiality. The laws governing the process of knowledge are just as dialectical as material processes. Knowledge, moreover, arises in a process of confrontation between the subject and objective reality" (ARRAZOLA & THIOLLENT, 2014).

The authors systematize environmental education as an instrument that needs to move towards transformations that strengthen sustainability through the harmony of nature with the basic needs of man, the conservation of life and the diversity of the planet, to guarantee a healthy environment and the needs of socio-economic and technological development. It is also responsible for getting people to reflect on their actions caused by consumerism and to promote changes in ideas and cultural values, to make the common good prevail over the individual, to make people know and demand their rights and to make them fulfill their obligations.

"The Brazilian National Environmental Education Policy, Law No. 9.795/99, approved in 1999 and regulated in 2002, highlights the development of educational actions and practices aimed at raising public awareness, organizing society in favor of environmental issues and defending the quality of life and the environment" (CARVALHO & OLIVEIRA, 2010).

The proposal brings together knowledge in the environmental area, especially degraded areas, with an emphasis on the recovery of springs through the use of consolidated technologies in riparian forest spaces and disturbed areas through the planting of seedlings of forest essences, fruit trees and creeping legumes, and will involve the participation of scholarship students, volunteers from IFPA - Campus Itaituba and the local community.

Moura & Barbosa (2013) state that "work projects are those developed by students, under the guidance of the teacher, with the aim of learning concepts and developing specific skills and abilities".

CHAPTER V

5. Area of study

5.1. Location of the Study Area

The research was carried out on the properties of small family farmers belonging to the Community Association of Sao Joao Batista do Curral Redondo and the area of the former municipal solid waste dump, in the west of the state of Parà.

The area has a well-defined history in relation to its origin and is directly related to the process of disordered urbanization that has taken place in recent decades in the municipality of Itaituba.

The area being surveyed is documented, belonging to the community that will now exercise legal rights (see figures 1 and 2).

Figure 1 - Location of the Municipality of Itaituba in the State of Parà

Location of the Sao Joao Batista Community Association in the municipality of Itaituba.

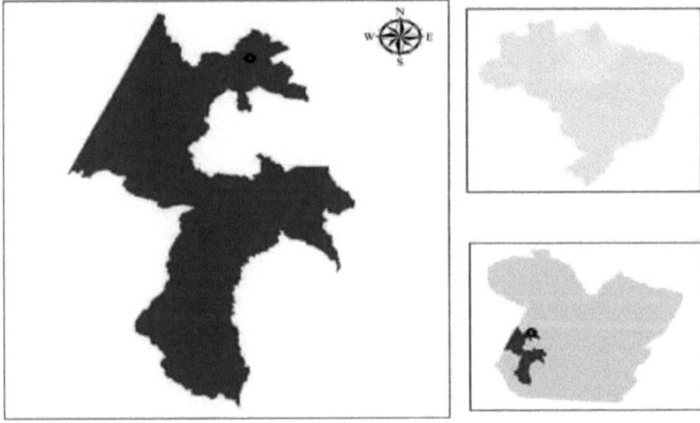

Source: Brian Almeida (2016)

Figure 2 - Location of the Study Area

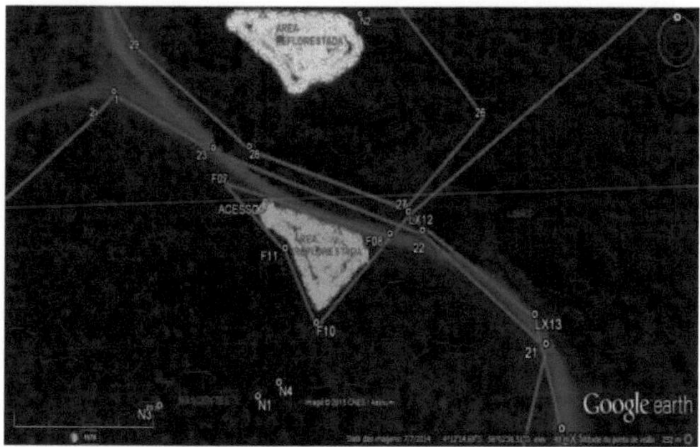

Source: Google/Liz Pereira (2015)

5.2. History of the area

The community covers an area of 299 hectares and is located on the municipal road that connects Itaituba to the district of Barreiras, starting at Km 09, which is accessible all year round. According to the residents, the area belonged to a businessman from the city, Mr. Arquimedes Mesquita, who had acquired the property from Congressman Davi Alves. It is an area that has been invaded by so-called "landless" people, most of whom have already sold to other landowners. The way the plots are organized is 50 meters in front by 100 meters in back and there are currently 40 families working with agriculture, small and large scale livestock. The community has a municipal elementary school which operates in the afternoon and morning shifts. It has two compartments where the classroom, the teaching support area and a leisure area are located, and it has a trained teacher, a pedagogue, who works with children aged between 6 and 14. Figure 3.

Figure 3: School in the Sao Joao Batista Community of Curral Redondo

Source: Researcher's collection (2015)

The community also has two religious entities: the Catholic Church and the Evangelical Assembly of God; a community hut for social events and a soccer field for sports. The main events in the community take place when the community needs to raise funds for actions that can improve the way it is organized, mainly soccer tournaments and religious festivities that take place in accordance with the plans of each religious entity.

CHAPTER VI

6. Methodological aspects

The research was based on a bibliographical survey with a qualitative and quantitative approach, using the field research method and participant research, applying the following techniques: semi-structured questionnaire, interview, observation at the research site, cross-sectional walk (MARCONI & LAKATOS, 2013), carrying out an experiment with seedlings of common Açai, Paricà and Fava Arara, to analyze the results in relation to the amount of macronutrients absorbed by them in a space of time of up to 4 months and building a dam 8 meters long by 4 meters wide to improve the environment of the springs.

The nursery was built on the grounds of the IFPA - Campus Itaituba, occupying a space of 300 square meters. It was supported by bamboo, smooth wire and 50% shade. Wooden beds were built measuring 1.20m wide by 15.0 meters long, with a total capacity to receive up to 10,000 seedlings in polyethylene bags measuring 18cm by 27cm and 15cm by 25cm. Seedlings of Paricà (Schizolombium amazonicum), Fava arara (Parkia multijuga), Brazilian mahogany (Swietenia macrifylla king), Jatobà (Hymenea courbaril L.), Andiroba (Carapa Guianesis aubi), Açai comum (Euterpe oleracea M.) and Buriti (Mauritia flexuosa).

An experiment was carried out in the nursery area with 3 (three) different substrates in an area of 9 m^2 , with the species Paricà (Schizolombium amazonicum), Fava arara (Parkia multijuga) and Açai comum (Euterpe oleracea m.), using a completely randomized design, over a period of up to 4 months, in a 3X3 factorial scheme, with 4 replications, totaling 36 plots, where the absorption rates of macronutrients N-Nitrogen (%), P-Phosphorus (mg/kg) and K-Potassium (mg/kg) plus C-Carbon (%) and O.M.-Organic Matter (%) were evaluated.Organic Matter (%) by the plants. The black soil used to produce the seedlings and the 03 specific substrates were analyzed: black soil with farmyard manure (70% soil and 30% manure), black soil with poultry litter (70% soil and 30% poultry litter) and black soil with organic compost (70% soil and 30% compost). Other comparative analyses were carried out on the plants, such as: stem diameter, number of leaves, weight of fresh and dehydrated mass.

The field research involved producing a diagnosis of the community association, the area, the vegetation, the families, socio-economic data, the history of the dump, the relevant legislation on solid waste, public policies and its purpose was to catalog all the necessary information. *"Field research consists of observing facts and phenomena as they occur spontaneously, collecting data related to them and recording variables that are presumed to be relevant, in order to analyze them"* (MARCONI & LAKATOS, 2013) and required bibliographical research on the subject and techniques according to the nature of the research that involved recording the data and then *analyzing* it, represented in figure 4.

Figure 4 Interview with the Director of Public Cleaning for the Municipality of Itaituba

Source: Researcher's collection (2015)

Rocha (2004) discusses participatory research, emphasizing its uniqueness and its two major dimensions, which involve the relationship between theory and practice in the search for a dialectical interaction, which means working from the perspective of practice. Its fundamental element is to bring out popular wisdom and tradition in the sense that they offer us clues to the social crisis we are experiencing today, which has given us a close relationship with the physical and social environment for this work. Participatory research is part of practical research. For Rocha apud Demo (2000), for the purposes of systematization, according to figure 5.

Figure 5 Planting Seedlings in the Anthropized Area

Source: Researcher's collection (2015)

6.1. Research stages

6.1.1 Participatory Meeting

In order to socialize the project and plan the actions with the producers' association, we initially held a meeting in the community to discuss the issue and involve the members of the association in the proposal, with the aim of getting them involved and committed to the reality that was being presented. Bibliographic data on the subject was collected, various readings were carried out and an action plan was organized, which was initially established with *on-site* visits to the research environment and the community to identify the region, determine the area of operation and the procedures for identifying the native vegetation and delimiting the area. Subsequently, visits were made to degraded environments, conservation areas and natural recomposition areas, as shown in figure 6.

Figure 6 Walk in the Research Area

Source: Researcher's collection (2014)

After this stage, data was collected to guide the starting point for contextualizing the problems and their links to the different themes, such as: agricultural production system, regional socio-economic characteristics, socio-cultural characteristics and the history of the region and its environment.

Part of this phase was carried out in a closed environment and another part in the field, gathering a series of data and information that allowed us to move forward in pursuit of the established objectives. From then on, we began to work in partnership with the community and the Federal Institute of Parà Campus Itaituba, and later other partners began to take part in actions in the field, the area that was the subject of the research, as shown in figure 7.

Figure 7. Meeting with community members

Source: Researcher's collection (2014)

6.1.2 Community diagnosis

In order to produce the diagnosis, some basic instruments were used: a) primary historical sources (subjects and official records) b) secondary sources (bibliographies, maps, cartographic charts, etc.), with a view to providing a theoretical basis for the object of the research, where data was collected on the socio-economic aspects of the families and the biophysical aspects of the area that was the object of the research, with a view to collecting the data needed to adapt the actions based on what the research proposed. This phase saw the collaboration of 03 teachers from the IFPA - Campus Itaituba, 01 (one) physicist and 01 (one) chemist, (01) biologist, as well as the participation of 04 (four) scholarship students from the Integrated Technical Courses. There was also the participation of volunteer students and teachers.

The project also involved partners throughout the research and implementation process. The emphasis was on *"in loco"* activities, targeting rural producers who work with family farming in the area being researched. For this stage, research techniques such as semi-structured questionnaires, interviews, traverses and observations were used to gather the information needed to build this history (see figure 7).

Figure 7. Family farm worker

Source: Researcher's collection (2015)

6.1.3 Construction of the Nursery for Seedling Production

The nursery for producing seedlings of forestry and fruit trees was built on the grounds of the IFPA - Campus Itaituba, in a space of 300 square meters, duly structured to meet the demands of the construction, production, maintenance and monitoring process.

The initial support structure was bamboo and later sawn wood, smooth wire and 50% shade cloth. Wooden beds were built measuring 1.20m wide by 15.0 meters long. The total capacity of the nursery was to receive up to 10,000 seedlings in polyethylene bags measuring 18cm by 27cm and 15cm by 25cm and an experiment with different substrates measuring $9m^2$, which were used for sowing Paricà (Schizolombium amazonicum), Fava arara (Parkia multijuga), Brazilian mahogany (Swietenia macrifylla king).Jatobà (Hymenea courbaril L.), Andiroba (Carapa Guianesis aubi), Açai comum (Euterpe oleracea m.) and Buriti (Mauritia flexuosa), shown in figure 8.

Figure 8 Construction of the Nursery in the IFPA Campus Itaituba Area

Source: Researcher's collection (2015)

The genetic material was acquired from reputable suppliers AIMEX - Associação das Empresas Exportadoras de Madeira do Parà APP - Área de Preservaçao Permanente and EMBRAPA - Empresa Brasileira de Pesquisa Agropecuâria. The material used to build the nursery was donated by the Sao Joao Batista community and IFPA - Campus Itaituba and Campus Castanhal, with other partners being open to financial aid such as CNPq - National Council for Scientific and Technological Development and other entities during the project's implementation stages. Students from the IFPA - Itaituba Campus helped with the construction of the nursery.

Specific and appropriate machinery was used to build the 8m x 4m dam and the (150) meter road: 01 (one) crawler tractor and 01 (one) truck, provided by the Itaituba City Hall, and a backhoe hired by a member of the community in partnership with the IFPA - Itaituba Campus. Community members, students and teachers were involved in this stage.

6.1.4 . Seedling Production Experiment with Different Substrates

The research applied in the nursery used an inter-causal design - DIC, for a period of up to 4 months, in a factorial scheme of 3 (types of substrates) X 3 (plant species), with 4 replications, totaling 36 plots, with the following distribution of treatments: Factor 1, plant species: Paricà (Schizolombium amazonicum), Açai (Euterpe oleracea m.). and Fava Arara (Parkia multijuga) and Factor 2, types of substrates: Treatment 1. TerraPreta (100%), (control);

Treatment 2: Black earth (70%) and farmyard manure (30%);

Treatment 3: Black earth (70%) and poultry litter (30%);

Treatment 4: Black Earth (70%) and Compost - Foliage and Manure (30%)

The experiment occupied an area of 9 m^2 , built in a wooden nursery, covered with 50% shade, and with a constant irrigation system, observing the needs of the seedlings until the end of the experiment. The seedlings were placed on the ground in polyethylene bags measuring 18 cm by 27 cm, with 8 holes.

The treatments were distributed on the plots entirely randomly by drawing lots. The DIC had only the principles of chance and repetition, with no local control and therefore the repetitions were not organized into blocks (RODRIGUES, 2014). In the environment, these seedlings received some special care such as: homogeneous environmental conditions (soil fertility, soil texture, uniform water distribution, etc.). The data was organized in tables with the variables for the specified treatments, which were laid out in the field and statistically analyzed using DIC Analysis of Variance, with Sources of Variation and Degrees of Freedom; Mean Test and Tukey's Test at 5% probability using ASSISTAT statistical software. Figure 9 shows the operation.

Figure 9. Experiment data collection and fresh matter weighing

Source: Researcher's collection (2015)

6.1.4.1. Chemical Analysis of Substrates:

For the analysis of macronutrients (Nitrogen - N, Phosphorus - P, Potassium - K) plus (Sodium - Na, Calcium - Ca, Magnesium - Aluminum - Al, Organic Matter - OM and Hydrogen Potential (pH), a sample was taken from each substrate, according to the treatments specified above. The aim was to chemically characterize the substrates and the black soil used to produce the Paricà, Açai and Fava Arara seedlings in the treatments.

The laboratory chemical analysis that determined the macronutrient concentrations in the black soil and the three substrates was carried out at the EMBRAPA-Eastern Amazon laboratory. Available phosphorus and potassium were extracted using the Mehlich 1 extractor, with the former determined by colorimetry and the latter by flame photometry. Calcium and magnesium were extracted using a 1N KCl solution and the titration method (EMBRAPA, 2011).

6.1.4.2. Chemical Analysis of the Plant Part - Stem and Leaves:

The chemical analysis of the plant part in the laboratory, which determined the concentrations of macronutrients (N, P, K + C and MO), was carried out on the dry plant material at a temperature of 60 °C. The plant material for this analysis was extracted from the following plants: Paricà, Açai and Fava Arara seedlings aged 4 months.

To determine the macronutrient content, samples were taken from the aerial part of the seedlings, which were cut, weighed and packaged, washed thoroughly, dried in the shade, packaged in special paper bags and dehydrated in an electric oven for up to 72 hours at 6060°C. They were then crushed in an appropriate mill, packed in special containers, the appropriate notes were made to control the material and then they were sent to the Emilio Goeldi laboratory in Belèm Parà. The method used was TEDESCO (1995).

6.1.4.3. Process of Planting Seedlings in Degraded Areas of Springs

The seedlings of forest and fruit trees were planted at a spacing of 5 meters between plants and 5 meters between rows, in a pit. No organic or chemical material was added, just the black soil that made up the seedlings in the bag. The Açai and Buriti seedlings were planted in the area of influence of the springs and the forest essences were planted in the areas that were bare, totaling 3.0 hectares. The seedlings were planted in the field at an age of 4 to 6 months and were produced in polyethylene bags measuring 18 cm X 27 cm and 15 cm X 25 cm, with 8 holes. The legume seeds were planted at a spacing of 0.80 x 0.50 cm and established in three specific areas of the springs which had been previously georeferenced. They were monitored in the field to check mortality and growth rates.

The area is part of the micro-basin of the "Passa Tudo" stream, which runs through a large part of the community. Forest and fruit plants were planted in the area with the aim of re-establishing the flora of the degraded environment, where they will have a direct impact on the relationship between soil and water. According to Trevisan (2010), "the soil-water interaction is influenced by the conditions of the soil surface, particularly its cover and the capacity of the soil to infiltrate water", particularly its cover and the soil's ability to infiltrate water", which means that rainwater running over bare soil is a source of sediment runoff, and is therefore liable to contaminate environments, including the water table.

The planting system obeyed the following specific groups. The two groups were planted in soils with different physical characteristics, but with the same essential and fruit trees: a) Paricà (Schizolombium amazonicum), Fava arara (Parkia multijuga), Mogno brasileiro (Swietenia macrifylla king), Jatobà (Hymenea courbaril L.), Andiroba (Carapa Guianesis aubi), Açai comum (Euterpe oleracea M.) and Buriti (Mauritia flexuosa).

Common açai (Euterpe oleracea m.) was also sown along the course of the springs and leguminous plants were planted: Feijao de porco (Canavalia ensiformis), feijao guandu (Cajanus cajan) and Mucuna preta (Mucuna pruriens). b) Paricà (Schizolombium amazonicum), Fava arara (Parkia multijuga), Mogno brasileiro (Swietenia macrifylla king).Jatobà (Hymenea courbaril L.), Andiroba, common açai (Euterpe oleracea M.) and Buriti (Mauritia flexuosa), seeds of common açai (Euterpe oleracea M.) in the course of the springs and sowing of legumes: Feijao de porco, feijao guandu and Mucuna preta.

Two forestry systems were organized in accordance with the technical criteria recommended for seedling production by official bodies: EMBRAPA - Brazilian Agricultural Research Corporation - CPATU and AIMEX - Association of Wood Exporting Companies of Parà APP - Permanent Preservation Area, IPAM - Amazon Environmental Research Institute, SEMMA's - Municipal Environment Secretariats, INPA - National Institute for Amazon Research, Ministry of Agriculture, Universities, Extension Services and Experiences of local and external Farmers, of which we list the use of seeds of provenance, selected seedlings produced in the project nursery. *"The disposal of seedlings must comply with the criteria of age, health, control, etc., as well as the monitoring of seedlings of native species* (PREGELLE et al. 2008).

These elements are of fundamental importance, not only for vegetative development but also for the incidence of pests and diseases. It is also important to establish special standards for preventive measures, with a view to minimizing future problems, of which we list a few: interspersed and diversified planting, using different families or different species so that there is not too much competition and, at the same time, trying to bring the vegetative development of the degraded area closer to that of the previous ecosystem. For this, technologies are essential in order to know, understand and apply them. For example, the SAF is a cultivation system that has shown good results in production activities, precisely because it is a diversified agroforestry system that performs differently in anthropized environments, as shown in figure 10.

Figure 10 Planting Açai, Jatobà, Paricà, Andiroba, Mahogany and Macaw Bean seedlings

Source: Researcher's collection (2015)

The planting of legumes in the recovery area consisted of the species feijao de porco, feijao guandu and mucuna preta. They were planted as a way of providing the soil with nitrogen. "Legumes are great nitrogen fixers" (BEGON, 2007), as shown in figure 11.

Figure 11 Planting Legumes, Pork Beans

Source: Researcher's collection (2014)

6.1.4.4. Continuous Monitoring

The purpose of the monitoring was to assess the mortality rate and vegetative development of the seedlings in the field, as well as to evaluate the germination rate of the açai and leguminous plants. In order to collect data to evaluate the results, the initial growth of the plants included in the forest restoration was monitored, and the water quality of the springs was observed: Ph and temperature. The forest species were observed, taking into account native species, selected according to their successional categories and their adaptability characteristics to the conditions of springs and/or riparian forest (Pioneers and non-pioneers) (PREGELLI et al. 2008) and alternatives based on results already experienced in the region with positive responses such as Mahogany, Paricà, Fava Arara, Jatobâ, Andiroba and others. This monitoring took place after planting to check the mortality rate. *(...) "The concern with repairing damage caused by man to ecosystems is not recent and is currently emerging as an important challenge to be overcome"* (CURY & CARVALHO JR. 2011).

CHAPTER VI I

7. Biophysical and Land Aspects of the Area

7.1. The St. John the Baptist community/association of "curral redondo"

The characterization of the research in the municipality of Itaituba, West Para, with family farmers from the Association of Producers of the Sao Joao Batista community of "Curral Redondo", arises from local demand. The presence of an open-air municipal garbage dump, deactivated within the community area, occupying an area of more than 22 hectares, compromising the hydrographic basin of the Igarapé "Passa Tudo", a stream that cuts through a large part of the territory and which, under the current circumstances, does not offer bathing water conditions in accordance with CONAMA Resolution 274/2000, due to the presence of thermotolerant coliforms, which, according to SEMAM/IEC data, are much higher than those considered "unfit". This compromises the daily dynamics of these populations (GOMES, 2014).

The Sao Joao Batista association stands out because of the layout of its activities and the way in which it is characterized. Its most recent history is centred on associations geared towards meeting the demands of small rural properties, organized in small farms, mainly developing family agriculture, with no standard of organized production systems. They are small farms producing beans, rice and cassava, mainly for food security and surplus. The produce is sold at street markets. The farms have fruit trees: açai, cupuaçu, orange, soursop, lime and jackfruit, small vegetable gardens, poultry and in some cases cattle breeding on small areas of cultivated pasture (brachiaria).

The community's economic support is directly related to production activities that have a degrading impact on the soil, mainly due to the practice of burning. *"Anthropic action has caused many changes to the soil, very quickly, not allowing it to recover*" "fires cause serious consequences for the environment, fauna and flora (MOTA 2010 & SOUSA et al 2015).

Another interesting aspect that can be seen in the community is the presence of landowners who have acquired plots of land through purchase, participate in the community, but their productive activities are not characterized as their main activity, some are small traders in the community and others in the city, in the most diverse branches of economic activity and maintain within the plot the figure of the hired caretaker, who earns up to 01 monthly minimum wage. It's worth noting that these spaces are characterized as areas for leisure at the weekends, although they do have some livestock and agricultural activities to support the family and the surplus goes to the market and the boss. Figure 12 shows the characteristics of anthropization.

Figure 12 Area of Capoeira Queimada

Source: Researcher's collection (2015)

In addition to the aspects *already* mentioned, there were moments of struggle in the community, mobilizing with the Federal Public Prosecutor's Office to intervene in the degrading situation, which culminated after much struggle in the suspension of the waste dump used by the municipality as a place to store urban waste. Some residents' accounts of the moment emphasize that "we managed to get the suspension and we were all mobilized to fence off the area, but in the late afternoon of that day the judge ordered them to continue, the municipality to continue depositing waste here". "We demobilized the group and went again to the Public Prosecutor's Office which, within a maximum of 30 days, reportedly announced the presence of the Federal Police and ordered the arrest of the municipal manager," and only in this way, and by agreement, was the dump closed.

For Toledo (2014), *"in the study of environmental sciences, various disciplines can be included in contributing solutions to a specific problem"*. In this interdisciplinary action, the relationship between Biology, Physics, Chemistry, Ecology, Geography, Agronomy, etc. is very clear, sciences that intertwine, complementing and systematizing concrete results, internalizing meanings to the dialectical processes of the propositions that take place in the daily life of modern societies, which is reinforced by Caporal (2006), when he refers to the science of agroecology, conceptualizing an interdisciplinary versatility, arising from each line of knowledge that complements each other, strengthening the prism of sustainability and the balance of the ecosystem in a specific environment.

7.2. Family Farming as a Basis for Environmental, Economic and Social Sustainability

Contextualizing family farming in the community leads us to reflect on the elementary historical concepts of transition in the processes that make up their relationship with the environment in which they live and their memories. According to Peterson (2009) "the contemporary world is going through an unprecedented crisis". Conceptualizing the exhaustion of a civilizational project, which is based on the accumulation of wealth in the hands of minorities and in fact, when projecting "the combination of the growing and increasingly urbanized world population, with the accelerated degradation of natural resources and global climate change, shapes a disturbing scenario that confronts us with decisive dilemmas".

Thus, looking at the context of the unbridled expansion of food empires, the context of recampesinization can be understood in terms of its quantitative dimension, when it refers to the increase in the number of peasant families and the democratization of the agrarian structure, and its qualitative dimension, considering the strengthening of the peasant nature in the part of family farming that has appropriated the entrepreneurial elements of production as a result of public access policies reaffirmed by modernization. For Pleog (2009), "the process of recampesinization cannot be confused at all with a return to the past". "In this way, as well as situating the presence of family farming in the historical process, the notion of recampesinization shows that the meaning of this process is not unidirectional, as the heralds of modernization proclaim."

In this direction, it can be seen that the productive activities of the families in the Sao Joao Batista community are manifested through the peasant actions of modernity, with their own objectives aimed at exploiting the eminently natural resources that permeate the dynamics of traditional environments. It has a system of organization that is legally established as a result of a statutory regime established in the context of relations with the urban area as a way of maintaining a business policy on the various products that come from the capitalist mode of production, reaffirms that "peasant family farming plays an important role in society and is centered on the struggle for autonomy and progress as a way of building and reproducing a rural livelihood in an adverse context characterized by relations of dependence, marginalization and deprivation" (see figure 13).

Figure 13 Area of Consortium Cultivation of Beans, Maize and Bananas

Source: Researcher's collection (2015)

This understanding is perceptible in the daily life of the community, where the presence of family farming is characterized as the main basis for survival, but concentrates a voracious desire for growth and transformation, aiming to find the path to social, economic and environmental autonomy, since these instruments are consolidated every year, The main focus is on environmental reconstruction and the growth of economic activities for family sustainability in the inhospitable and degrading environment caused by the environmental impacts caused by the presence of solid urban waste, improperly deposited by the municipal government over the years.

The concern with which they deal with diversified production is visible in the productive environments (farms). (...) "peasant production aims to reproduce, improve and expand ecological capital, produce marketable surpluses and create networks and institutional arrangements that enable both production and reproduction" (PLORG, 2009), as can be seen in figure 14.

Figure 14 Cassava and Bean Cultivation Area

Source: Researcher's collection (2015)

Peasant resistance is still the main driving force behind food production, and it is re-established even with the processes of modern societies. According to Wanderley (2009), "the practices of peasant resistance, their form of organization, their reproduction strategies, relations with the market, integration processes and others, are among the dimensions of peasant existence, the knowledge of which has been consolidated with the multiplication of theses and academic research and the constitution of new spaces for debate and reflection" (see figure 15).

Figure 15 Action by Public Bodies and Students in the Community, SAGRI and PMI

Source: Researcher's collection (2014)

In this relationship of existence and resistance, the community fights and struggles with the aim of conquering its space, transforming dialogical conceptions to the detriment of transformations that can reduce differences and minimize environmental problems, maximizing economic aspects through production surpluses and social aspects through the constitutional rights of citizenship and public policies implemented in recent years that have partly valued family farming. For example: encouraging agriculture through the constitutional programs FNO and PRONAF and social programs, the zero hunger program, PAC, rural education, housing credit, family grants and so on (NASCIMENTO, 2015), these actions have become important for the transformations that have taken place in the country in recent decades, representing a turning point in Brazil's historical trajectory.

A fundamental element that must not be overlooked is the new revolutionary policy that has been established for formal and informal education at all levels.

The strengthening of family farming is centered precisely on this conception of values that will give sustainability to the process of growth and development of the territory, conceptualized by the multiplicity of arrangements 70

production, verticalization of production, adding value, collective and organized marketing, an effective accounting control system, a healthy product and a good relationship between producer and consumer.

According to Ploeg (2009), "consumers increasingly value authenticity, freshly harvested products, taste and diversity and are willing to pay producers who are engaged in new and appropriate forms of sustainability".) And this represents a differentiated conception of the means of production, given the way in which the peasantry is articulated and fostered, and the social movements represent this own and articulated characteristic, aimed at the well-being of the less favored populations, which is complemented by the achievements of the system of modernization of Brazilian education, starting with Law 9394/96. (See figure 16).

Figure 16 Meeting with the Community Association

Source: Researcher's collection (2014)

Caporal (2006), Ploeg (2009) and Altieri (1995) make important points about agroecological issues and food security, as well as the need for forms of associative organization to organize production and strengthen verticalization, with a view to adding value and reorganizing the championship in the form of cooperation, with a view to competition and access to public policies for small rural entrepreneurs.

For the Brazilian exponent of education, Paulo Freire*, "education does not educate the world, education educates people and people educate the world".* This phrase shows us how transformative education is and how it plays a fundamental role in the life of modern societies.

The social organization acts in a social and religious way to define the actions pertinent to daily life, the relationship with public authorities, as well as the guidelines related to the development of productive activities programmed generally by month.

7.3. Land Structure of Plots in the Community

The rural properties that make up the community area have no legal documentation, the right to ownership is only that of possession, it was found that only 10 (ten) properties are in the process of being regularized, which have already been georeferenced, as one member of the board comments. *"Friendship is very important, because we went to the Terra Legal program people who were doing other work in the municipality and through a friend who contacted them, we were helped, the team demarcated these 10 plots, it was the people who believed in the work done by the board".* There are also, according to reports, individual owners who are doing the demarcation with their own resources, those who don't depend on the property to generate income, but only as a leisure space.

The relevant legislation places responsibility for this issue with the State Division for Land Regularization in the Legal Amazon (SRFA) (INCRA, 2011).

In this direction, what we have realized throughout this research is that the land situation is a major problem for access to land and public policies, since historically it has been observed that the western part of the state of Parà, especially the municipalities that make up the Tapajós micro-region, is still an isolated space, insipient from the presence of public power.

The dynamic we are witnessing at the moment is the application of Law No. 13.001/2014, one of whose fundamental characteristics is the issue of control. *"The environmental regularity of agrarian reform projects, through the inclusion of settlements in the Rural Environmental Registry (CAR), as a control mechanism and conditions for financial regularization"* (MDA, 2014),

Even if we look at the law and the criteria it establishes, especially with regard to technical assistance, the presence of these organizations in the community is practically absent. We can see that family farmers don't receive any support to strengthen these guidelines for the implementation of constitutional principles with regard to access to land, property and public policies, with a view to living as citizens. See figure 17.

Figure 17 Map of the Anthropized Area

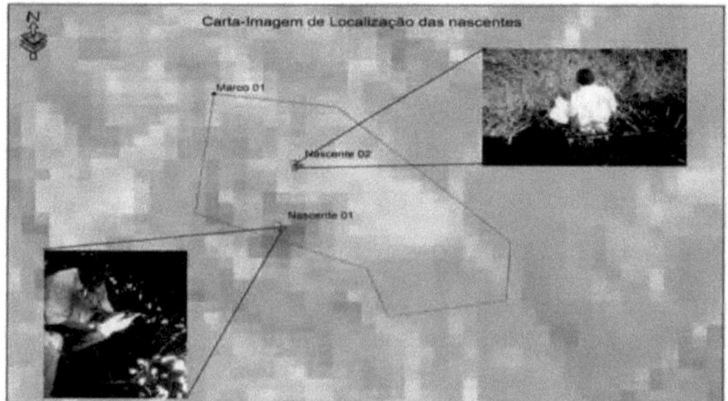

Source: Researcher's collection and Brian (2014)

7.4. Characterization and Land Use

The predominant characteristics of the soil in the community, the area that was the subject of the research, after on-site observations and bibliographic references, are yellow and red latosols, lateritic concretionary soils and nitosols. "Soils correspond to the living layer that covers the surface of the earth, in permanent evolution, through the alteration of rocks and pedogenetic processes commanded by physical, biological and chemical agents" (LORENZO, 2010), the soil is very important for ecosystems, as it plays a fundamental role in the development of the various forms of life on the planet. It has different physical and chemical characteristics in different environments (see figure 18).

Figure 18 Soils of the Area in the Community

Source: Researcher's collection (2015).

In our observations, even though the soil has a special characteristic for such a situation, it can be seen that over time various actions have been carried out in the area, directly damaging the soil and surface and underground water sources, due to the incorrect disposal of solid waste in the open. According to Mota (2010), *"the control of solid waste pollution is achieved by avoiding its incorrect disposal". "When it rains, leachate is washed away along with other materials by torrents, reaching rivers and lakes. Leachate can also reach groundwater"*, says CAVINATTO (2013). This fact is very present in the environment, compromising the dynamics of the natural environment, totally altering the ecosystem and harming an entire population.

Given this situation, we can see that there are other relevant situations of soil contamination, such as the unpleasant aesthetic aspects, the disfigurement of the landscape, the unpleasant smell, the proliferation of insects and rodents, which transmit diseases, the production of gases, mainly methane and others. Soil pollution from garbage can result in air pollution when waste is burned in the open (IBIDEN). "Animals proliferate in these deposits, spreading microbes that transmit serious diseases, such as leptospirosis, caused by a type of bacteria found in rat urine" CAVINATTO (2013). For all these reasons, we understand that waste should be disposed of correctly in landfills, incineration and composting plants, as shown in figure 19.

Figure 19 Solid waste disposal areas

Source: Community collection (2011)

The way it was used is characterized by the opening of large cells that received large quantities of material that were buried without the slightest care, easily leading to water permeability, causing a large production of leachate, discharged in the winter period by underground waters forming large environments conducive to runoff to the springs and the "Passa Tudo" stream, which has high levels of pollution, (SEMMAM/IEC 2003).

Characterizing and assessing the potential of natural resources is an important guiding tool for decision-making and land use planning. Predicting and determining the suitability of these resources will establish guidelines for different uses, which are the basic and fundamental elements for any definition of sustainable development (GAMA et al., 2007).

7.4.1 The Use of Fire and the Consequences for the Soil

The community found that some properties still use fire, mainly in pasture areas. According to Cury & Carvalho Jr. (2011), "fire has always been used as a necessary tool for maintaining agricultural and livestock fields".

However, this is still the only alternative that makes it feasible to prepare land for annual cultivation due to the low purchasing power of the families, who are unable to receive technological aid due to the conditions in which they find themselves and also due to the absence of public authorities.

In this field, some properties stand out for their use of the soil with some differentiated activities and do not use the resource of fire, as some agro-ecological activities of cultivation and breeding are being applied, using natural resources to protect springs, planting forest essences and fruit trees and promoting aspects of agro-forestry. Others are raising fish rationally on a small scale. In this way, *"a water and soil management strategy is created, encouraging the organization and participation of producers, the community and governmental and non-governmental institutes" (SEMA, 2010)*. "The scope and relevance of these issues show why the study of land use in its various forms has become one of the most active areas of interdisciplinary research in Brazil and the world over the last two decades" (TOLEDO, 2014). Figure 20

Figure 20 Area in the process of recovery

Source: Researcher's collection (2015)

1.5. Vegetation

Specific to the region, the community is characterized by open tropical forest, a hot and humid climate, an average annual temperature of 26.7°C and an annual rainfall of 2,189.2mm (INMET, 2013). For the community members, the history of the flora in the area that was the subject of the research was characterized by "closed" forest, virgin forest, which has been transformed over the years, especially since the 1980s, when the Federal Government's National Integration Programs were implemented.

Farmers' testimonies:

Interviewee "a"

"In the 80s, this was just "closed" forest, now there is little forest, there is a lot of capoeira and pasture, there is only virgin forest there in Mr. Juarez". Area bordering the community.

Interviewee "b"

Mr. Cearà and Tarraca "there were a lot of trees here, he mentions Mahogany (Swietenia macrophilla king), Ipê (Tabebuia chrysotricha), Sapucaia (Lecythis pisonis), Freijó (Cordia goeldiana Huber), Castanheira (Bertholletia excelsa), Itaùba (Mezilaurus *itauba*), Cedar (Cedrela odorata L.), Angelin (Dinizia excelsa Ducke), Tatajuba (Bagassa guianensis Aub), Inajà (Attalea Maripa or Maximiliana), Gameleira (Ficus adhatodifolia), Barriguda (Chorisia glaziovii), Cumaru (Dipteryx odorata), Uxi (Endopleura uchi), Cajui (Anacardium Humile), Copaiba (Copaifera langsdorfii), Jatoba (Hymenaea courbari) and Andiroba (Carapa guianensis)".

Interviewee "c"

It refers to fruit plants such as Cajà (Spondias lutea L), Jenipapo (Genipa americana), Macaùba (Acrocomia aculeata), Tucuma (Astrocaryum aculeatum), Buriti (Mauritia flexuosa), Cajù (Anacardium occidentale), Goiaba (Psidium guajava), Murici (Byrsonima crassifolia L. Rich), Mutamba (Guazuma ulmifolia Lam), Bacaba (Oenocarpus bacaba), Açai (Euterpe olearacea) and other palm trees".

Based on the information collected, it can be seen that today the natural landscape has undergone very strong intervention, since there is very little evidence of primary vegetation and very small patches of secondary capoeirao vegetation. We note this statement: "*Madeireira Itaituba was very present in the region and exploited a lot of wood here*".

Nowadays, the areas are small, with limited areas of no more than 1.0 to 5.0 hectares. Some of these areas maintain small secondary reserves and some producers apply agro-ecological techniques, planting and preserving, mainly the springs and streams, but the streams are compromised by the high concentration of products silted up by the rains, as shown in figure 21.

Figure 21 Protected area

Source: Researcher's collection (2015)

1.6. Water Resources

The main water resources in the community are represented by the micro-basin of the "Passa Tudo" stream, which belongs to the Tapajós River Basin. The results presented show that the waters of the igarapé are compromised and unsuitable for human consumption and bathing, according to the recommendations of CONAMA 274/2000, due to the pollution caused by the leachate coming from the subsoil of the dump and via surface water, especially during the rainy season (SEMMAM/IEC/2003). This will require further research and analysis in the future (see figure 22).

Figure 22 Community water resources

Source: Researcher's collection (2014)

There is a great wealth of springs in the community that form a fundamental component in the dynamics of nature, participating in practically all life cycles and human activities on the planet (SEMA-PR, 2010).

What can be observed in the protected areas known as APPs in the environment visited is that several properties do not comply with the relevant legislation regarding the preservation of riparian forests, rivers, streams, springs, streams, etc. (CFB, 2012), because deforestation compromises the existing populations in the ecosystem.

The Department of the Environment of Paranà - SEMA-PR (2010) uses this item as a concept, to which this paragraph refers, emphasizing that: *"the protection of riparian areas to springs and rivers through the reestablishment of forest in preservation areas and in other areas of rural properties are fundamental in the protection and production of water"*.

In recent times, water resources have become a global concern and a cause of conflict, caused by the scarcity of drinking water and climate change, and mainly by anthropogenic actions (SCHONS, et. Al, 2014). (...) *"In a rural ecosystem, the set of agricultural activities can be responsible for significant changes in the primitive environment"* PHILIPPI Jr. (2005). In this way, it can be seen that the impacts on water resources are worrying, in terms of both qualitative and quantitative aspects.

There is also other striking evidence in rural areas, in view of the viability of production, especially when chemical products such as fertilizers and pesticides are used, making producers responsible for environmental contamination and damage. "The accelerated growth of cities has caused wildlife to migrate from forest and river areas to live ever closer to human beings" (SOUZA, 2014). It is clear that this closer relationship is generating major problems that unbalance life forms in the natural ecosystem, accelerating this damage.

1.7. Environmental degradation

The area that was the subject of the research has a high level of environmental degradation, given the condition of the soil, surface water and groundwater from the springs. Visually, the water has a different color and a strong odor during the rainy season, so it is believed that this change is caused by the presence of waste discharged into the water, from surface water, organic matter and slurry discharged into the Igarapé. The natural vegetation has been completely altered and the resilience of the native species is low. It is important to note that the environment is compromised, according to previous SEMMAM/IEC (2003) analyses.

Nowadays, even with the closure of the dump, we find a lot of material in the process of decomposition, plastics, hardware and glass. "Disposal in inappropriate places visibly degrades the landscape, whether urban or rural. Open-air dumps produce polluting gases, which influence the dreaded greenhouse effect, and stink. They also encourage the presence of animals and disease vectors" (SOUSA, et al. 2014 p : 31). Today, this situation is no longer the case, but there is a lot of material in the area that is in the process of being degraded and some that is not, mainly glass, tires, hardware and various plastics such as pet bottles and bags in general (see figure 23).

Figure 23 Anthropized area of the Old Dumpsite

Source: Researcher's collection (2014)

The final disposal of municipal solid waste (MSW) has become a major environmental problem, since municipalities all over the country have encountered major problems, even those that already carry out effective recycling actions and operate landfills, because what is known as waste remains, a source of environmental pollution of considerable proportions (PIRETE et al., 2014).

Over the years, industrial and technological development, intense production activity and high consumption have led to more and more problems that have led to, mitigated and inflated other environments to unprecedented levels of degradation (...) "*poor* solid waste management causes environmental impacts and compromises public health" (OLIVEIRA et al., 2014).

These are constant challenges in the daily lives of modern societies in search of alternative solutions capable of minimizing impacts on the planet, but the actions proposed are still incipient and are not moving in the same direction as the global world and its needs. Federal Law No. 12.305, passed on August 2, 2010, provides elements that will allow states and municipalities to behave differently when it comes to the production, collection, treatment and final disposal of their waste.

CHAPTER VII I

8. Socio-economic Aspects of the Farmers of the Sao Joao Batista Association

8.1. Social and economic organization of the community

The socio-economic aspects of the community show three groups that differ in terms of their economic activity. There are members of the community who make their living directly from production; there are those who carry out other commercial activities in the community or in the municipality's headquarters, the urban area, keeping a caretaker; and others who work on properties outside the community, but maintain their subsistence farming activities. The origins of these producers are diverse: Goiàs, Rio Grande do Norte, Maranhao, Porto Velho and Parà, from the cities of Monte Alegre, Itaituba, Marabà and Alenquer. Sacramenta & Costa (s/r) emphasize this moment in our history by saying that *"there is an intense process of migration, with people moving from all over the country as a result of the policies being developed to populate the Amazon"*.

Associativism as a form of social organization is characterized as an alternative for the community's demands and aspirations vis-à-vis public authorities. Meetings are held in the community hut. The community has a Catholic and an Evangelical church (Assembléia de Deus). The community school is a municipal school with a primary level of education, serving the entire community, with one teacher serving the students in the early grades. They promote sports and festive activities when necessary, to strengthen actions within the community. There are notorious situations of conflicting interests between certain internal groups that often differ or do not share the same ideas. The electoral political issue is still a dividing line for these issues, interfering directly in the daily life of the community, causing losses in the participation of its members in relevant community actions. As for the presence of public authorities, the residents interviewed said that municipal health services are present through the health agent and periodic vaccination campaigns.

The presence of ATER services is non-existent, and when they do occur, it is only for specific projects via credit from official financial institutions. They have already received PRONAF for the second time. Their economic activities are characterized by family subsistence production. For Andrade, according to Cavalcante 2014, *"The long period of man/nature toil in the history of this village has been responsible for the framework of human evolution based mainly on the abundant products of the forest"* and what can be seen is that modern society increasingly demands natural resources, which serve the industry of unbridled and marginal consumption, with a view to keeping the ills of savage capitalism warm.

The community's economic activities also include raising free-range poultry, fish, small vegetable gardens and extensive cattle rearing. These are the basis of the community's production, which they sell at the town's main market.

8.2. The environment and recovery from the impacts of anthropization

The restoration of degraded areas is based on specific issues that meet the specific requirements and recommendations of CONAMA. It involves an environmental diagnostic assessment, with a view to ascertaining the state of degradation and aims to define the most appropriate restoration technique. In this way, it provides the researcher or producer and others with the most correct definition to be applied.

This leads us to define the principle of forest recovery or restoration. According to Martins (2007), "Restoration is the return of a degraded ecosystem or wild population to an undegraded condition, which may be different from its original condition".

A degraded area is one that has suffered, to some degree, disturbances to its integrity, whether of a physical, chemical or biological nature. Recovery, in turn, is the reversion of a degraded condition to a non-degraded condition, so recovery can be understood as a set of actions necessary for the area to be fit again for some productive use in conditions of environmental balance" (MARTINS, 2007). Its objectives should be to recover its physical, chemical and biological integrity (structure) and, at the same time, to recover its productive capacity (function), whether in the production of food and raw materials or in the provision of environmental services. Adopted, via the planting of wild forest seedlings, equivalent to the region's natural ecosystem; Planting of fruit and leguminous plant seedlings.

In this sense, according to the nature and severity of the degradation, as well as the effort required to reverse this state, the following cases can be considered: (see tables 01 and 02).

Rehabilitation: returning the degraded area to an intermediate state of the original condition. In this case there is a need for anthropogenic intervention.

Redefinition or redestination: recovery of the area with a view to a different use/destination from the pre-existing situation, with the need for strong anthropogenic intervention.

Table 01: Forest rehabilitation via seedling planting

Planting technique	Advantages	Disadvantages	Recommendations
Row planting	Rapid soil coverage. Less weeding maintenance. Reduced maintenance costs.	High implementation costs. Difficulties in obtaining seedlings.	Close to seedling nurseries. Availability of different species. Availability of human and financial resources.

Source: IPAM Forest Restoration Manual, 2011.

Table 02: Rehabilitation with Legume Seeds

Planting technique	Advantages	Disadvantages	Recommendations
Manual planting	Can be associated with agroforestry	Larger workforce starts	Difficult to access, small areas.

Source: IPAM Forest Restoration Manual, 2011.

The description in the tables took us on the road to implementing the field practices, where all the possibilities were analyzed so that they best suited our reality. In this way, line and manual planting was carried out in the degraded area, following the previously determined spacing.

It should be noted that after these decisions, the seedlings performed well at first due to the rains, but over time they became compromised due to lack of water. The recovery of the area is in a slow process due to a series of factors involving financial resources and labor capacity over the course of the week.

Given the events that have taken place and the slow pace of future results, it is believed that a much better response will be possible in 2016. According to Cury and Carvalho Jr. (2011*). "the areas that can be restored are the Permanent Preservation Areas" (APPs)*, because they perform an environmental function of preserving water resources, the landscape and geological stability, the balance of the ecosystem as a whole in terms of soil protection, fauna and flora (see figure 24).

Figure 24 - Area being reclaimed by planting seedlings

Source: Researcher's collection (2014)

Cury and Carvalho Jr. (2011*)* consider APPs (permanent preservation areas) to be forests and other forms of natural vegetation located along rivers or any watercourse from its highest level in a marginal strip, the minimum width of which will be the description referred to in Brazilian legislation establishing defined parameters for such issues.

The banks of the APP, in order to protect and prevent water pollution, popularly known as water pollution, caused by industries and also by people, especially in urban areas, who throw waste into lakes and streams, causing flooding problems, compromising the aesthetics of the city, putting people's lives at risk by increasing the rodent population and the proliferation of diseases.

Since the reformulation of the Brazilian Forest Code, a series of changes have been made to the protection of the banks of rivers and other watercourses, as a way of protecting and preserving areas of riparian forest, as shown in table 03, with regard to preventive measures for APPs.

Table 03: Marginal strip with minimum width for APPs

PPA width	River width
30 m wide	Up to 10m
50 m wide	Between 10 and 50 m
100 m wide	Between 50 and 200 m
200 m wide	Between 200 and 600 m
500m wide	Larger than 600 m

Source: adapted from the Manual for Forest Restoration - IPAM - 2011.

In order to analyze the quality of the surface water of the springs and the "Passa Tudo" stream, some very specific criteria were adopted as a preponderant factor for human health, established by the Water Quality Index (IQA), using as a standard source those adopted by the Environmental Company of the State of São Paulo (CETESB, 2004). The following table shows the IQA classification indices.

Table 04: IQA Classification - (Water Quality Index)

Track	Classification
80 - 100	Great
52 - 79	Good
37 - 51	Acceptable
20 - 36	Suitable for conventional treatment
0 - 19	Inappropriate

Source: CETESB (2004)

It should be noted that the two springs are directly influenced by the area where the deactivated solid waste dump is located, which generates leachate that spreads into the soil, contaminating the watercourses that flow into the igarapé, promoting pollution and compromising the entire natural ecosystem.

A number of important criteria were taken into account when planting the forest species: Identification of the degradation factors to isolate the area; Assessment of the history of degradation and land use; Identification of the potential for natural regeneration and the most appropriate forest restoration techniques for the degraded APP based on the analyses carried out (CURY & CARVALHO Jr, 2011).

According to EMBRAPA (2008), reforestation with tree species native to the Amazon brings with it a series of indispensable components for balancing the ecosystem and is closely related to improving soil fertility. Planting legumes (figure 25) has the ability to fix nitrogen from the atmosphere (N_2), in addition to environmental benefits such as erosion control, better use of rainwater and reduced contamination of watercourses by fertilizers, as noted by authors such as (BEGON, 2007).

Figure 25 Potential Area for Natural Regeneration

Source: Researcher's collection (2015)

CHAPTER I X

9. Results and Discussion

9.1. Planting seedlings

The seedlings planted play a very important role in protecting the soil and the springs, which are in the process of vegetative development. It is noticeable that, as a result of the drought, there was a stunting in the development of some plants and, in particular, the Açai, Buriti and Fava Arara seedlings were the ones that had problems with the mortality rate, up to 100%. As for the Paricà, Jatobà and Andiroba forest species, the mortality rate was less pronounced, at around 67%. With replanting, the results improved, especially the legumes (Figure 26).

Figure 26 Planting Shrub Legumes to Protect the Soil

Source: Researcher's collection (2015).

It is worth noting that the Açai seeds that were planted on the banks of the riparian vegetation were the ones that showed the best results, as they continued to grow slowly due to the lack of suitable climatic conditions, especially in relation to the rise in temperature and the absence of water during the period. *"Permanent soil cover by plants or plant residues, in addition to containing surface runoff, favors the infiltration of water into the soil, minimizes water loss through evaporation and helps purify it, given the filtering capacity of the soil"* (SEMA-PR, 2010).

Although the environment is understood today as the set of natural resources and their interrelationships with living beings, it is common to see this concept associated only with the "green" of the landscape, with nature or with life (SMA/SP, 2006). When you relate these aspects to what is proposed in the field, based on these interventions, you realize that this action is much more than the quote emphasizes. (see figure 27).

Figure 27 Prepared area, planting seedlings of forest essences and legumes.

Source: Researcher's collection (2015)

Planting in a line and at a spacing of 5 meters has brought great benefits, as it facilitates mechanized maintenance and provides opportunities for the use of other pioneer and non-pioneer forest species and even the possibility of developing agroforestry systems in the environment that can be organized in accordance with local production arrangements. This phase provided an opportunity for theory and practice with students from the Environmental Sanitation course and community members. The spacing adopted was determined for all the forest species as shown in figure 28.

Figure 28 Plant spacing

X	X	X	X	X	X	X
X	X	X	X	X	X	X
X	X	X	X	X	X	X
X	X	X	X	X	X	X
X	X	X	x 5__	X	X	X 5

Source: Researcher's collection (2015)

9.2. Analysis of Vegetative Development and Protection of Springs

The seedlings were planted in the field at an average height of: Paricà (Schizolombium amazonicum), 0.80 cm tall, Fava arara (Parkia multijuga), 0.75 cm tall, Brazilian mahogany (Swietenia macrifylla king), 0.60 cm tall, Jatobà (Hymenea courbaril L.), 0.60 cm high, Andiroba (Carapa Guianesis aubi), 0.60 cm high, Açai comum (Euterpe oleracea m.), 30 cm high, and Buriti (Mauritia flexuosa) 30 cm high, and were planted in January and February

2015, after 7 months of age. The planting system for the seedlings followed special parameters, using a spacing of 5m X 5m between rows and plants. Due to the characteristics of the soil, which is very clayey, the initial pits did not follow the standard size of 0.40cm/0.40cm/0.40cm, but the size of 0.30cm/0.30cm/0.30cm. No chemical or organic fertilizer was used in the holes at this stage. In the second phase of planting, farmyard manure was used, averaging 1 kg/hole. We observed the best performance and resistance to dry weather of Paricà, Mahogany, Jatobà, andiroba and the legumes Feijao de porco and guandu. Black mucuna did not germinate in the environment.

It is believed that climatic factors such as the lack of rainfall during the period and high temperatures caused the death and inhibition of the vegetative development of the seedlings in the first year, but there was a good recovery the following year. During the period, we also found that the springs in 2014 only appeared in December, between the 10th and 20th of the same month, while in 2015, until the end of December, there was no water in the springs. Today, we find a very significant volume of water after 3 years of activity, as shown in figure 29.

Figure 29 Spring area

Source: Researcher's collection (2014).

According to the Brazilian Forestry Code, "riparian forests bring several benefits, such as ensuring soil stability, preventing soil erosion and landslides, preventing solid particles, pollutants and residues such as pesticides from being carried into watercourses, causing contamination and siltation. Tree canopies cushion the impact of rainwater on the soil, preventing it from becoming compacted, and guaranteeing food for fish and other aquatic animals". For Cury & Carvalho Jr. (2011), it helps to maintain stable water temperatures due to the climate formed under the tree canopies, connects forest fragments, "forming corridors" that serve as a refuge for wild animals, prevents water scarcity and ensures long-lasting, clean water sources for consumption.

In this way, the ecosystem and the life that exists there are kept in balance. "Biodiversity conservation, climate change and, above all, the new model of sustainable development have been fundamental in recognizing the importance and urgency with which environmental issues must be addressed" (SMA/SP, 2006). The first spring stands out because it is the first to emerge in the environment and has good water potential during the rainy season between December and June. There are two other springs in this environment that only appear at the end of January, as shown in figure 30.

Figure 30 Area of the Outcropped Spring in the Process of Recovery, Initial Containment

Source: Researcher's collection (2014/2015)

The observation made is that upwelling is directly related to rainfall. If there is a lot of rainfall, the rocks emerge earlier. However, if rainfall is scarce, it alters the period of upwelling. Three containment barriers were built, one made of tires with wood, another with wood and tubing and a dam using clay, wood and tires with dimensions of 8 meters long, 5 meters wide at the base and 4 meters wide at the surface and 2 meters high, as shown in figure 31.

Figure 31 Containment dam

Source: Researcher's collection (2015)

The construction of the containment dam to protect and maintain the water source that will supply the environment during critical periods of the year took several stages: The dam was built using some of the technologies adopted in the reduced impact logging system (IFT, 2009). Choosing the most appropriate environment for opening the road, using only one tractor blade for the width, slope, least impact on the vegetation, best space for building the dam and observing the area's natural resources. It was made with suitable soil from the site itself (Yellow and Red Latosol), as shown in figure 32:

Figure 32 Springs Area, Final Containment Dam

Source: Researcher's collection (2015)

A flow system was used with two PVC pipes, each 100 mm long, and a drainage system (sluice gate) in case of excess water during the rainy season. A bulldozer and a backhoe, wood, pipes, nails, a chainsaw and a hammer were used for the construction. Teachers, scholarship students, volunteers, community members, trainee students and partners all took part.

"Among the areas that can be restored are the Permanent Preservation Areas (PPAs). An APP is any strip of land covered by natural vegetation and which has the environmental function of preserving water resources, the landscape, geological stability, biodiversity, the gene flow of fauna and flora, with the aim of protecting the soil and the well-being of the population, within the limits defined by current legislation" (CURY & Carvalho Jr. 2011).

9.3. Seedling Production Experiment with Different Substrates

9.3.1 Results of the Chemical Analysis of the Soil and Substrates

For Brasil & Cravo (2007), "the interpretation of the results of the analysis

involves an agronomic and economic evaluation of the relationship between the nutrient content of the soil obtained by an analytical method and the response to the fertilizer".

Table 5 shows the results of the soil analysis, based on the samples collected, and gives the following interpretation: The chemical element Nitrogen in the substrate black earth with poultry litter was the most representative with 0.14%, Organic Matter showed results with small variations between the samples.

We can also see that the black earth/compost substrate stood out, with 16.75g.kg^{-1} . Phosphorus also stood out in poultry litter, compost and farmyard manure, more so than in black earth, with indices higher than 30 mg.dm^{-3} , considered to be of high fertility (IBIDEN);

Potassium showed significant results in the poultry litter and compost substrates, with indices above 90 mg.dm^{-3} , considered to be highly fertile. According to the indices presented in this regard, the black soil and farmyard manure were below 45 mg.dm^{-3} , considered to be of low fertility.

Looking at the table, it can be seen that the pH of the substrates - poultry litter, compost and farmyard manure - has a good hydrogen potential, above 7.0. In this way, the presence of aluminum in the substrates was totally neutral, as shown in Table 5.

Table 5: Results of Chemical Analysis of Soil and Substrates

Identification of treatments	**N**	**MO**	**PH**	**P**	**K**	**In**	**Ca**	**Ca+Mg**	**Al**
	%	g.kg^{-1}	Water	---- mg.dm^{-3}			cmolc.dm^{-3} ---------		
Black earth	0,04	14,43	5,7	2	2	2	2,4	2,9	0,1
Black earth with poultry litter	0,14	15,82	7,3	584	239	210	2,4	5,1	0,0
Black soil with compost	0,05	16,75	7,3	41	292	4	3,0	4,5	0,0
Black soil with farmyard manure	0,05	15,49	7,4	43	18	5	3,0	3,9	0,0

Source: EMBRAPA CPATU (24/12/2014)

In the process of recommending correctives and fertilizers, it is essential to use the results of soil analysis as efficiently as possible. However, the technician who advises the farmer on these practices does not always make use of all the information that can be obtained from the soil analysis report, often underutilizing fundamental aspects for decision-making (LOPES & GUILHERME. 2004). To this end, we tried to establish minimum conditions for the production of seedlings in the experiment that would justify the result, with the aim of applying it in the environment and in other areas. The discussion of the experiments dealt with this nutrient/plant relationship and the effects of fertilization on the plant tissues of the seedlings in the different substrates.

9.3.2 . Results of Plant Foliar Chemical Analysis of Seedlings

According to Kurihara apud Malavolta et al. (2005), "the assessment of the nutritional status of plants is usually done by leaf analysis, bearing in mind that the newly mature leaf is the organ that, as a general rule, responds best to variations in the supply of nutrients, whether from the soil or from fertilizer". He goes on to say that the main metabolic reactions and physiological changes resulting from nutritional disorders occur in the leaves.

According to Lopes (2004), a good interpretation of leaf analysis makes it possible to verify various situations and list nutrient deficiencies, toxicities or imbalances, making it possible to monitor and evaluate a fertilization program and its function is to reduce soil fertility problems.

Paricà seedlings: the species Schizolombium amazonicum, commonly used in different Brazilian states, is known as Paricà, but has other names (DUARTE, 2014, apud BARNEBY, 1996). In the nursery, the seedlings showed good vegetative development and the material was collected for leaf chemical analysis. According to Santos Filho et al. (2007), information on the nutritional requirements of forest species, especially native species, is scarce. He also notes that "the nutritional requirements of native species have been quite different", which can be seen in the results of the comparative analyses of Paricà in the different substrates.

An alternative source of organic matter is animal waste, the use of which in agriculture is an increasingly common practice, as it serves as a way of disposing of waste and cycling nutrients within the unit that generates the organic waste (VIEIRA & WEBER, 2013, apud SANTOS et al., 2007).

According to Souza et al. (2003), "(...) the recommended fertilization for the production of Paricà seedlings can be done with 2 to 3 kg per cubic meter of NPK substrate, formula 15-30-15 and liming with 6 kg of lime per cubic meter of substrate". That didn't happen with the substrate in the experiment, no chemical fertilizer was used.

The aim of soil analysis is to interpret the results in order to assess the levels of absorption by the plants in the nursery. The sufficiency band methodology can be used, with the comparative use of nutrient concentration tables in which it is possible to identify the nutrients that can affect crop yields, without, however, establishing the degree of nutrient impairment in planting nutrients and their effect on productivity. According to the authors, the study revealed that raising the pH of the area to 5.5 improved the availability of potassium and phosphorus. In addition, despite this, the levels of potassium applied did not significantly affect the growth in height and diameter of the species.

One of the characteristics of the experiment was that no chemical products were added, since we adopted agroecological principles for the production of seedlings using organic products, farmyard manure, poultry litter and organic compost.

Although a certain fraction of the organic matter in manure is decomposed and released over a period of one to two years, another fraction is transformed into humus, which is more stable. In this form, the nutrients are released slowly. Thus, the components of manure converted into humus will have a persistent and lasting influence on the soil (VIEIRA & WEBER, 2013 apud BRADY, 1989).

The results of the experiment showed that the Paricà in the specified substrates had the best vegetative growth in the nursery and in the planting area, i.e. in the field. It is being used as an organic green fertilizer for the soil. Studying the changes in the chemical properties of a dystrophic red latosol, subjected to treatments with organic waste, promoted significant changes.

Through these actions, the Paricà seedlings showed the following results:

Carbon (**C**), although the black soil with farmyard manure treatment stood out from the others, there was no significant difference between them;

Phosphorus (**P**) followed the same line, with the black composted soil treatment standing out, but there was no significant difference between them;

Potassium (**K**) gives us a different result, since the two treatments, black earth/ poultry litter and black earth/ compost, do not differ significantly from each other, but do differ significantly from the black earth/ farmyard manure and black earth treatments, respectively;

Nitrogen (**N**), although the black earth with poultry litter treatment stood out from the others, there was no significant difference between them.

The interpretative analysis of the results was carried out in order to better understand the dynamics that are directly related to the black earth soil used as a control and the specific substrates that were used to produce the seedlings in the nursery. When we make these approaches, we will realize that the nutritional specificities are quite significant in relation to the soil used. This can be seen in the following figures (see figure 33)

Figure 33 - Results of the leaf chemistry analysis of Paricà seedlings

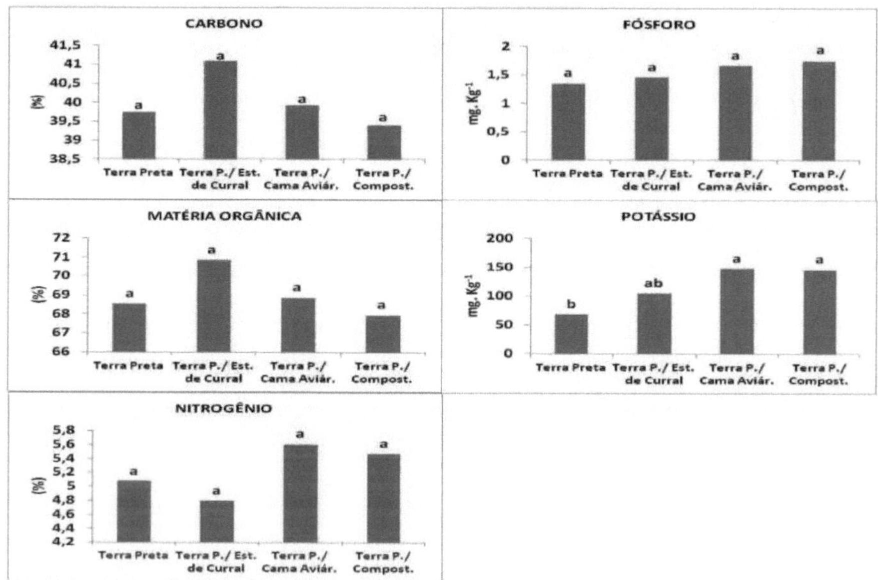

The results obtained from the leaf analysis of the seedlings themselves reflect these aspects, since the comparisons between the results of the analysis of the 03 substrates (Table 1) and the results of the experiment, shown through the leaf analysis, clearly show this perception, since the differences are established precisely because of the use of organic supplements. The substrates composed of 70% black earth and 30% poultry litter, black earth farmyard manure and black earth compost both stood out in relation to black earth alone, used as a control.

The macronutrients (N.P.K + MO and, C) stood out. However, Potassium showed no significant difference in the black soil with Paricâ treatment. This demonstrates the importance of using organic fertilizers in agroecological systems.

For Altieri (1995), "agroecology in the world incorporates ideas about focusing on agriculture that is linked to the environment and more socially sustainable, centered on the soil and production, and also on the production of ecologically sustainable systems". He emphasizes the broader meaning of the word, saying that "agroecology refers to the study of ecological phenomena" and is fundamental in view of the environmental foundations of the global world.

Açai seedlings: The açai tree occurs spontaneously in Brazil in the states of Amapà, Maranhao, Parà, Tocantins and Mato Grosso, (OLIVEIRA et al. s/d,). Breaking the Brazilian borders, it is found in other countries of the world such as Guyana, French Guiana, Suriname and others, but the Amazon concentrates the largest share of these species in the estuary of the Amazon River, where dense and diverse populations are found (IBIDEN).

Salman et al. (2008) state that "in the Amazon region there is a great diversity of tree species that are of great interest to man, both for economic exploitation and for restoring ecological balance". "Studies on the nutrition and fertilization of the açaizeiro are extremely incipient, and there are no consistent results that allow us to accurately assess the nutritional status of the plants and, above all, to establish fertilization recommendations" (OLIVEIRA et al., n.d.).

In the nursery, the seedlings performed well, but in some plots there were losses of vegetative material or they did not germinate satisfactorily. (...) "The nutritional requirements of native species have been quite different", which we will see in the results of the comparative analyses of the different substrates.

Through these actions, the Açai seedlings showed the following results (Figure 32):

Carbon (**C**), although the black earth with poultry litter treatment stood out from the others, there was no significant difference between them;

In terms of phosphorus (**P**), the black earth with poultry litter treatment was significantly superior to the others. There was no significant difference between the treatments black earth/urban manure, black earth/compost and black earth;

Organic Matter (**OM**) stands out more in the black earth with poultry litter treatment, but there was no significant difference between them;

Potassium (**K**) shows a different result, since the three treatments black earth/urban manure, black earth/ poultry litter and black earth compost did not differ significantly from each other, but were statistically superior to the treatment with black earth alone;

Nitrogen (**N**), although the black earth with poultry litter treatment stood out from the others, there was no significant difference between them, as shown in figure 34.

Figure 34 - Results of the Açai leaf chemistry analysis

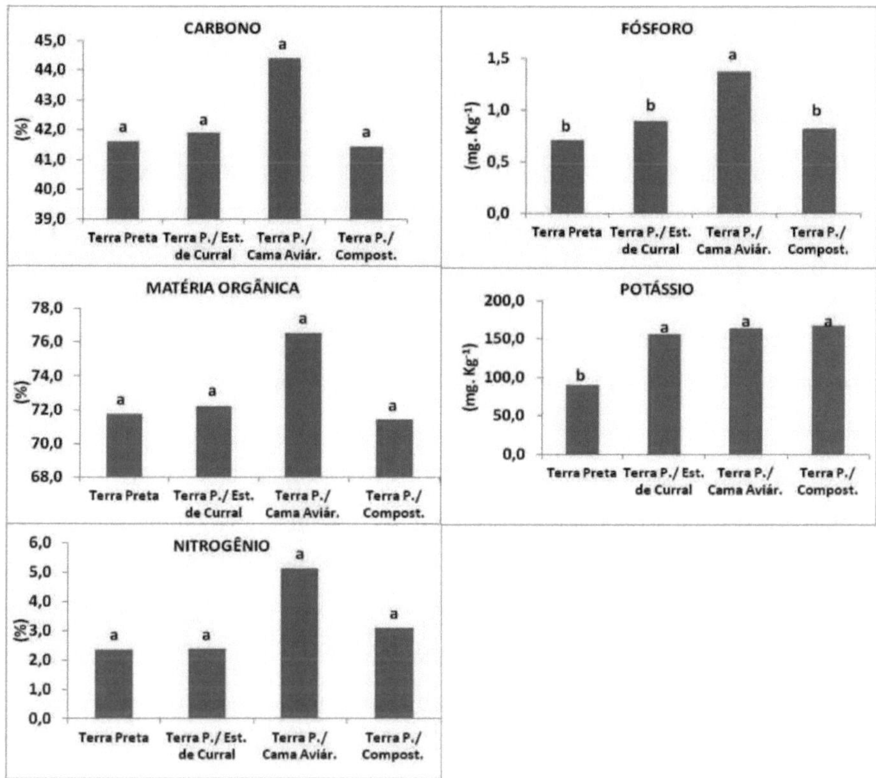

The results obtained from the leaf analysis of the seedlings themselves reflect these aspects, since the comparisons between the results of the analysis of the three substrates, as shown in Table 5, and the results of the experiment, demonstrated through the leaf analysis, make this perception clear, since the differences are established precisely because of the use of mineral supplements. The substrates composed of 70% black earth and 30% poultry litter, black earth farmyard manure and black earth compost, both stood out in relation to the black earth alone, used as a control. The phosphorus and potassium in the Açai, black earth and poultry litter treatments were the most significant.

The Açai material analyzed, taken from the nursery experiment, received the same water, light, temperature and cultivation treatments for a period of 4 months. There were no significant changes between the substrates, poultry litter, farmyard manure and organic compost, especially in terms of color and vegetative development, which took an average of 30 days to germinate. The control, treated with black soil, had a lower germination rate and the development of the seedlings was different, being described in terms of vegetative development. Dapont (2012) comments that "monocot seeds are typically albuminous and have hypogeous germination", i.e. those that develop in the soil.

Neto (2004, apud Dapont, 2012) emphasizes the theory about the germination process and seedling growth of the açaizeiro (*Euterpe oleràcea*) "that the projections of the ruminated forehead of the seed seem to play a role in transporting sugars produced from the reserves of the endosperm to feed the germinating embryo". Oliveira et al. (s/d), emphasizes that "the fact that the seeds do not germinate and the seedlings paralyze or reduce growth in an anoxic environment (absence of oxygen), explains the lower frequency of the species in permanently flooded areas".

The use of leaf analysis as a diagnostic criterion is based on the premise that there is a relationship between the supply of nutrients in the soil and their levels in plants, and that increases or decreases in levels are related to higher or lower yields (IBIDEN). However, the relationship between the levels of a given nutrient and dry matter production may not be as simple or as straightforward as this definition suggests.

<u>Fava Arara seedlings:</u> "this is a forest species belonging to the Mimosideata legume family and its habitat is the terra firme and vàrzea forests on clay soil in the Amazon region" (ROCHA et al., (1999). In the experiment, the seeds were sown in polyethylene plastic bags, with the substrate being 70% black soil + 30% substrates (farmyard manure, poultry litter and organic compost).

Salman et al. (2008) state that "in the Amazon region there is a great diversity of tree species that are of great interest to man, both for economic exploitation and for restoring ecological balance". "Studies on the nutrition and fertilization of the Fava Arara are extremely incipient, and there are no consistent results that allow us to accurately assess the nutritional status of the plants and, above all, to establish fertilization recommendations" (OLIVEIRA et al., n.d.). In the nursery, 15% of the seedlings had problems with germination. There was a situation where the whole plot was lost or the number of seedlings per plot was reduced by up to 89%. Each plot had 10 seedlings (...) "the nutritional requirements of native species have been quite different", which we will see in the results of the comparative analyses of the different substrates.

The results obtained from the leaf analysis of the Fava Arara seedlings are reflected in the elementary concepts established, since the comparisons between the results of the substrate analysis, seen in table 5, of the analysis of the soil and substrate, is evidenced in the results of the experiment.

This is clear because the differences are probably due to the use of organic supplements. Thus, leaf analysis as a diagnostic criterion is based on the premise that there is a soil-plant relationship depending on the supply of nutrients in the soil and those absorbed by the plants (IBIDEN). However, the relationship between the levels of a given nutrient and dry matter production may not be as simple or direct as this definition suggests.

Through these actions, the Fava Arara seedlings showed the following results.

Carbon (**C**), The black earth/ poultry litter treatment stood out from the others, statistically there was no significant difference between the treatments for this parameter;

Phosphorus (**P**) The black earth/composting treatment stood out from the others; statistically, there was no significant difference between the treatments for this parameter;

Organic Matter (**OM**) stood out in the black earth/ poultry litter treatment, but there was no significant difference between them;

Potassium (**K**) gives us a different result, since the three treatments black earth/urban manure, black earth/ poultry litter and black earth/composting did not differ significantly from each other, but were considered significantly superior to the black earth treatment;

Nitrogen (**N**), the black earth/composting treatment stood out from the others, there was no significant difference between the treatments, statistically superior to black earth, shown in figure 35.

Figure 35 - Result of the leaf chemistry analysis of the Macaw Bean

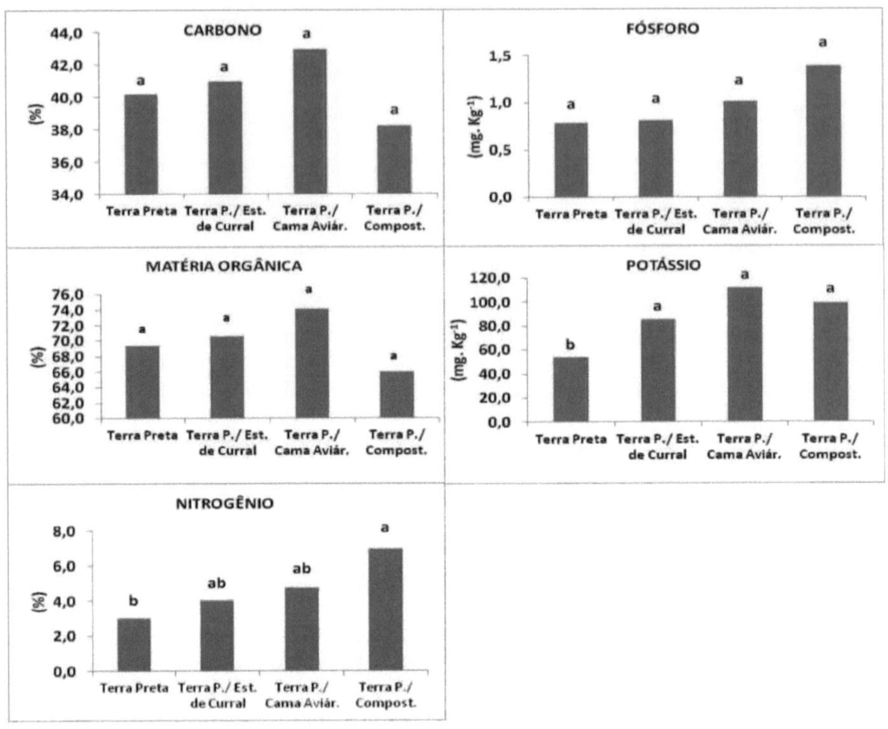

9.3.3. Results of the seedlings' plant performance

PARICA

Fresh mass: According to Souza et al. apud Rossi and Quisen (2003), "the growth of Paricà is quite rapid", which was noticeable in the nursery with the seedlings. In the definitive planting area, the vegetative development of the Paricà showed a proportional development between the seedlings on average 0.90 cm in height, however, some plants developed more quickly than others which is probably related to the characteristics of the soil, since the most suitable for this type of cultivation has a less clayey characteristic "(IBIDEN). This is not the case in the research area, because the soils are quite clayey. According to Rossa et al. (2013) "Paricà plants showed higher survival rates in the first, sixth and twelfth months after planting, with survival rates of 98.75, 91.72 and 96.67% in the monoculture system, respectively".

According to Rossa et al. (2013), "Seedlings with low quality standards show lower growth increments (...)". The paricà showed a high volume of green mass, but there was a considerable reduction in the volume of dry mass. Therefore, it can be seen that there is a volume of concentrated water that is lost in relation, for example, to the water present in the fresh mass of the Açai and, analyzing the dry mass, this situation shows the differences between them, in the treatments with substrates, For Rondon (2002), he makes some observations about biomass production, highlighting that spacing is a determining factor for this production and that there are variations depending on the species and the age of the plants.

Dry mass: It can be seen that the ratio of green mass to dry mass, Figure 34, shows equal results, the treatment black earth/urban manure and black earth/ poultry litter, showed that there was no significant difference between the two, but in relation to black earth and compost there was a significant difference and, considering the analysis of the compost substrate in relation to black earth there was also a significant difference, "substrates intended for seedling production must be fertile and free of pathogens, in order to favor germination and seedling development" (ROSSA et al. 2013).

Diameter: There were no significant differences between the treatments in absolute values. The poultry litter and compost treatments stood out compared to the manure and black soil treatments, but there was no significant difference between them. The author conceptualizes the diameter of the plants in relation to the increase in population, showing that an increase in population density leads to a reduction in plant height and diameter. In the nursery this concept is different, the number of populations was the same for the treatments and the same spacing in the different substrates and the results showed no significant differences. See Figure 35.

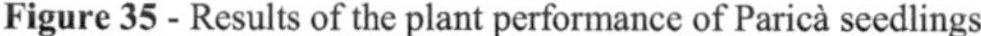
Figure 35 - Results of the plant performance of Paricà seedlings

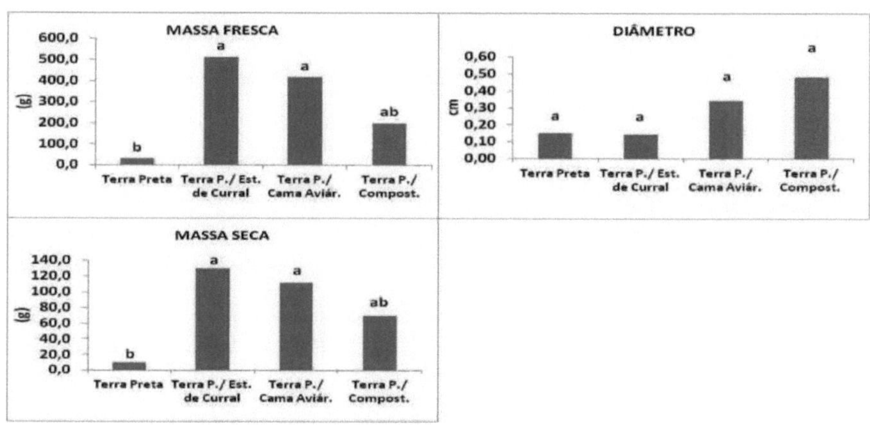

HONEY

Fresh mass: The results of the fresh mass performance of the Açai seedlings stand out as follows: the black earth/urban manure and black earth/ poultry litter treatments were not statistically different from each other and were considered superior to the black earth/compost and black earth treatments respectively.

Unfortunately, it was not possible to collect field data on the açaizeiro due to the failure of the seedlings in the field. They were planted according to basic technical principles. However, they were unable to withstand the severe drought and high temperatures in the region, which meant that they had to be replanted again, because ecologically, these palm trees play an important role in the riparian vegetation ecosystem. "(...) the oligarchic nature of this forest is determined by the flooding regime, since a small number of tree species have adaptive mechanisms to survive in soils with low oxygen tension. In the case of the açaizeiro, these mechanisms are represented by morphological and anatomical adaptations, such as: roots that emerge from the rootstock above the soil surface, the presence of lenticels and aerenchyma in the roots (OLIVEIRA et al.,s⁄d).

The Açai seedlings used for the samples taken from the experiment with substrate had two leaves, a diameter of between 0.25mm and 0.3mm and a height of between 0.25cm and 30cm, while the smaller ones had diameters of 0.20mm on average and a height of between 0.20cm and 0.25cm with two leaves. This shows the importance of organic fertilizer for seedling growth:

Dry mass: It can be seen that the treatments of farmyard manure, poultry litter and compost stand out, but statistically they don't show significant results, while there was a significant difference in relation to black earth. An interesting aspect to consider is the amount of water present in the fresh material of the Açai, as the results of the plant analysis show that there was little loss of water, with no significant differences. However, when compared to Paricà, the levels of water loss are much lower.

Diameter: There was no statistically significant difference between the treatments, as shown in figure 36.

Figure 36 - Results of the plant performance of açai seedlings

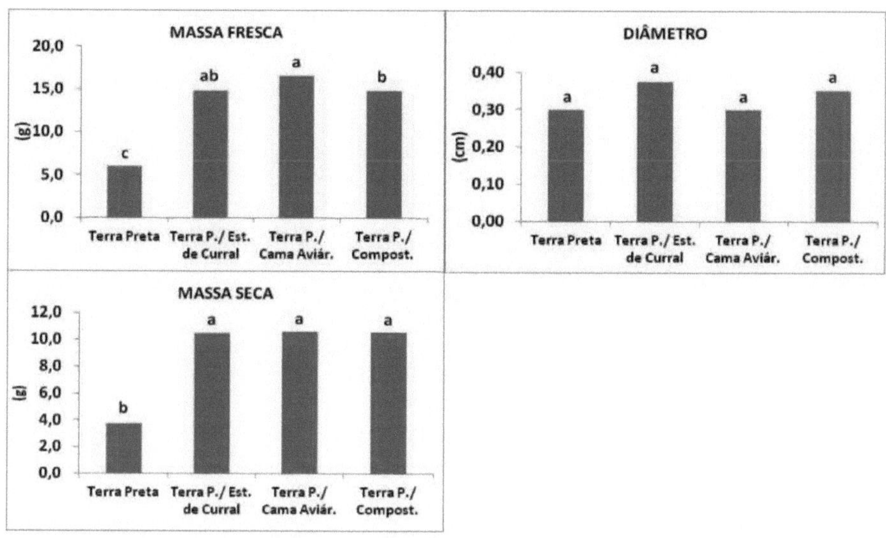

FAVA ARARA

Fresh mass: The results of the fresh mass performance of the Fava Arara seedlings stand out as follows: the black earth/urban manure treatment showed a significant difference compared to black earth and poultry litter, which did not occur with composting, as there was no significant difference between the two treatments.

Dry mass: The results reflected the green mass, the performance of the black earth/urban manure treatment showed a significant difference compared to black earth and poultry litter, which did not occur with compost, as there was no significant difference between the two treatments.

Diameter: The performance of the black earth/urban manure treatment showed a significant difference in relation to black earth and poultry litter, which did not occur with composting, as there was no significant difference between the two treatments, according to

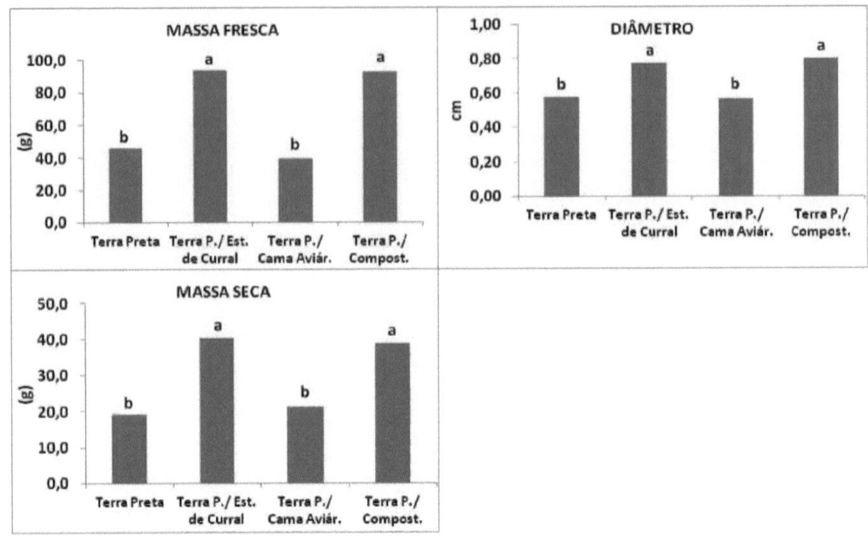

The Fava Arara seedlings in the nursery used for the samples taken from the substrate experiment had two leaves, a diameter of between 0.48mm and 0.60mm and a height of between 0.68cm and 0.78cm. The control, treated with black soil, had a lower germination rate and the seedlings developed differently, with smaller diameters of 0.45mm on average and heights of 0.42 cm to 0.68 cm, with 3 pairs of leaves. This shows the importance of organic fertilizer for seedling growth.

11. Final considerations

In recent times, urbanization combined with the consumption of industrialized, non-durable and disposable products has led to a large increase in the volume and variety of solid waste generated in urban systems, posing challenges for public managers.

Today, in the middle of the 21st century, the Amazon, as a peripheral frontier of contemporary capitalism, is a strategic territorial space for the world due to its biodiversity stocks, abundant water resources and climate services, which are essential for the maintenance and balance of the planet.

Itaituba, according to the Director of Public Cleaning, produces an average of 50 tons of urban waste a day and the aggravating factor is that there is no selective collection, only a group of waste pickers from an Association collects it from the dump for recycling, emphasizing that in the future this organization, together with the City Hall, will develop in partnership a proposal to improve this reality.

Disorderly urbanization has had a serious impact on the environment, affecting flora, fauna, terrain, water sources and the climate, causing significant changes, resulting in pollution and degradation, as well as other problems such as noise pollution, visual pollution, water, soil and air pollution, sewage and the production of large volumes of waste.

The presence of solid waste dumped for decades in the community area of the São Joao Batista Association of Curral Redondo has considerably compromised the water sources in the Igarapé Passa Tudo, Bom Jardim and the ecosystem as a whole. As such, it is unleashing an unprecedented environmental crime, requiring effective intervention by society and the public authorities, at all levels of government.

When I started the work, I didn't realize how big the problem was. It was challenging, exhausting, and I still had to live with the memory of those who had lived through degrading times there. The days went by and the challenges only increased, trying to understand the daily dynamics of these populations and their challenges.

It is believed that the measures taken for the rehabilitation or redefinition process, i.e. the return of the degraded area to an intermediate state of the original condition, in these cases under anthropic intervention, as the proposed objectives and floristic recomposition are fully established, the environment of the springs and the ecosystem as a whole will be re-established.

The production of the diagnosis reveals how much the community needs technical information on cultivation and financial support from public policies; organizational support and essential services in the areas of health, education, side roads and technical assistance.

The challenges are many, there is still a lot to be done and fortunately the proposal for its continuity has been understood. A new goal has been set, a period of up to 8 years of research, which will certainly be strengthened by other professionals and other institutes in the future.

It should be noted that this research is of an initial nature and raises other opportunities and challenges, enabling new work and new results, new goals that are not so simple or easy to achieve, but it is understood that much more can be done. In the municipality, anthropic action extends to other spaces that degrade as society does not participate in the process of transformation, and in this regard, education is fundamental.

The work presents some preliminary data, which may be altered as it is not conclusive, as it requires further studies in search of other answers.

REFERENCES

ABNT- Brazilian Association of Technical Standards. IFPA Academic Work Standardization Manual - 2015 - 2020.

ABNT- Brazilian Association of Technical Standards. Lutheran University of Brazil Martinho Lutero Library/Canoas. Resources sector, 2014

ALTIERI, M. A. Agroecologia Bases cientificas para uma agricultura sustentable, Editorial Nordan-Comunidad 1995

ARAÙJO, M. P. Serviço de limpeza urbana à luz da lei de saneamento bàsico: **regulaçâo juridica e concessâo da disposiçâo final de lixo** / Marcos Paulo Marques Araùjo, prefacio Luiz Oliveira Castro Jungstedt, Belo Horizonte: Forum, 2008.

ARRAZOLA, L. S. D. & **THIOLLENT,** M. J. M.(Org.). **Methodology, Theory of Knowledge and Action Research** - UFPA, 2014, p.54.

AMAZON BIOME. Brazilian Forestry Institute. In' www.ibflorestas.org.br/bioma-amazônico.html, 2015.

BEGON, M. Ecology : **from individual to ecosystem** / Michael Begon, Colin R. Townsend, John L. Harper; translated by Adriano Sanches Melo ...(et al.). - 4ª ed. - Porto Alegre : Armed, 2007. 752 p. : ill.: 28cm.

BRAGA, B. et al. **Introduçâo a Engenharia Ambiental** - 2a. ed. Sao Paulo : Pearson Prentice Hall, 2005. p. 2-9.

BRASIL. Código Florestal e Legislaçâo Correlata - 2nd ed. -Brasilia : Senado Federal, Subsecretaria de Ediçôes Técnicas, 2010. 156 p. (Environmental Collection, v. 4)

BRAZIL, Ministry of Agrarian Development (MDA), National Institute for Colonization and Agrarian Reform, Santarém Regional Superintendence: **Annual ordinary management accounts for 2010,** INCRA, 2011.

BRAZIL, Sustainable Amazon Plan (PAS), 2006.

BRAZIL, Sustainable BR-163 Plan, 2006.

BRAZIL, Ministry of the Environment 2006.

BRAZIL. Ministry of Education. Coordination for the Improvement of Higher Education

Personnel - Capes. **Contribution of Brazilian postgraduate courses to sustainable development:** Capes at Rio + 20 / Brasilia : Capes, 2012. 194, p.

BRUSEKE, F. J. Desenvolvimento e natureza: **estudos para uma sociedade sustentâvel /** Clovis Cavalcante, organizador - 5 ed. - Sao Paulo : Cortêz ; Recife, PE : Fundaçao Joaquim Nabuco, 2009.

CAPORAL, F. R. et al. Agroecology: **disciplinary matrix or new paradigm for sustainable rural development.** Brasilia - 2006.

CARVALHO, A. R. & **OLIVEIRA,** M. V. C. **Principios bàsicos de Saneamento do Meio,** 10ª ed. Sao Paulo : ed. Senac Sao Paulo, 2010.

CASTRO, C. et al., Soybean production technologies - central region of Brazil 2009 and 2010. Londrina: Embrapa Soja, 2008. 261 p. (Embrapa Soja. Sistemas de Produçao, 13), site https://www.agencia.cnptia.embrapa.br

CAVALCANTE, A. E. C. Technological extension in the use of organic waste: **its transformation into fertilizer for the sustainability of agriculture in the municipality of Irituia, Parà**, 2014, p.48.

CAVINATTO, V. M. Sanitation: **a source of health and well-being.** 2nd ed.

Sao Paulo. Moderna, 2013.

COELHO, C. M. Passagens. Revista Internacional de História Politica e Cultura Juridica Rio de Janeiro: vol. 5, no.3, 2013.

CURY R. T. S. & **CARVALHO JR.** O. Manual for forest restoration in transitional forests. **Good practices series:** Canarana 2001.

CRUZ, E. D. & **PEREIRA,** A. G. Technical Manual, Seed Germination of Amazonian Species:Fava-Arara-Tucupi (*Parkiamultijuga* Benth.) June, 2015 Belém, PA, 2015.

DAPONT, L. C., Dissertation: Acceleration of Germination and Shading in the Formation of Açai Seedlings. Rio Branco, 2012.

DERISIO, J. C. **Introduçâo ao controle de poluiçâo ambientalali'** ed atual . Sao Paulo: oficina de Textos, 2012.

DEMO, P. Qualitative research. Finding a balance between form and content. Rev.latino-am.enfermagem, Ribeirao Preto, v. 6. n 2. p. 89-104, 1998.

DEMO, P. Pesquisa participante. **Myth and reality**, Rio de Janeiro, SENAC/DN, 1984. p. 112.

DIAS, G. F. Mudança Climàtica e Você - Cenârios, **desafios, Governança, Oportunidades, Cinismos e Maluquices**, ed. Sao Paulo : Gaia, 2014.

DUARTE, D. M. Growth of Paricà seedlings subjected to defoliation and water deficit, Ipameri, 2014.

EMBRAPA. Serviço Nacional de Levantamento e Conservaçao de Solos (Rio de Janeiro, RJ). **Manual of Soil Analysis Methods**. 2 ed. Rio de Janeiro, 1997.

EMBRAPA. II International Symposium on Tropical Savannas: **Recovering springs in cerrado areas,** EMBRAPA gado de corte, Campo Grande, Brazil. 2008.

BAPTIST FACULTY OF MINAS GERAIS: Center for Extension, Postgraduate Studies and Research - CEPPES. **Manual de Normatizaçao da Faculdade Batista de Minas Gerais, 2014/1.**

FERREIRA, C. P. **Nutrientes do Solo em Sistemas Agricolas na Microrregiâo de Castanhal, Parâ.**

FERREIRA, J. N. S. **História & Estórias e outras Moagens**. 1ª ed. Itaituba. Copyright, 2010.

GAIGER, L. I. **The solidarity economy in the face of the capitalist mode of production.** Caderno CRH, 16 (39), 181-211, 2003.

GOMES, F. A. & SILVA, J. P. Avaliaçao da balneabilidade da praia do sapo**,** In'**ALMEIDA,** Djalmira de Sà (Org.) **Produzindo textos acadêmicos**: saneamento bàsico, saùde, cultura, sociedade e meio ambiente sustentâvel no Tapajós. Curitiba - PR, 2015, p: 99.

GOUDIE, A. The Encyclopaedic Dictionary of Physical Geography**.** Oxford, England: Brasil Blackwell, 1995. 528, p.

GUERRA, A.J.T. A contribuição da geomorfologia no estudo dos recursos hydricos, **BAHIA ANÀLISE & DADOS** Salvador, v. n. ESPECIAL**,** p. 385389, 2003.

GUERRA, A.J.T. Encosta e a questão ambiental. In: CUNHA, S. B.; GUERRA, A. J. T. (Org.). **Questâo Ambiental: diferentes abordagens**. Rio de Janeiro: Brazil, 1999. P.191-218.

GUERRA, A.J.T. & **CUNHA,** Sandra Baptista da. Impactos Ambientais Urbanos no Brasil - 10 ed. - Rio de Janeiro: Bertrand Brasil, 2001.

IBAMA, Brazilian Institute of Renewable Natural Resources, 1990.

ICMBio, Management Plan for the Itaituba I and Itaituba II National Forests, 2014.

JESUS, R. P. de. Associativism in Brazil and societies in the 19th century: **a critical repertoire of Council of State records (1860 - 1889)**. Artigo, Juiz de Fora v. 13, n. I, p. 144-170, 2007.

KURIH ARA, C. H. et al. Interpreting Leaf Analysis Results, EMBRAPA, Embrapa Agropecuâria Oeste; Colombo; Embrapa Florestas, 2005.

Law No. 12.305/2010, on the National Solid Waste Policy.

LENTINI, M. et all, Fatos Florestais da Amazônia, Belém : IMAZON, 2003.

LEAL, J. W. et. al. **Primaz**: Mineral integration program in the municipality of Itaituba, 1996.

LOCATELLI, M. et al. Soil Chemical Characterization and Growth of Schyzolobium parahyba var. amazonicum (Huber ex Ducke) in Agroforestry, Agroecological Systems in the State of Rondônia 2012.

LOPES, A. S. & **GUILHERME,** L. R. G. **Interpreting Soil Analysis:** Concepts and Applications, ANDA, 2004.

LORENZI, H. Arvores brasileiras : **manual de identificaçào e cultivo de plantas arbóreas nativas do Brasil,** vol. 3 / Instituto Plantarum, 2009.

LORENZO, M. P. PEDOLOGIA - Concepts of soils and degraded areas, 2010. In' *marianaplorenzo.com/.../**pedologia-conceitos-de-solos-e-areas-** degradada...*

LUCKESI, C. C. & **PASSOS,** E. S. Introduçao à Filosofia: **Aprendendo a pensar -** 7ª ed. Sao Paulo: Cortez, 2012. p. 17-21

LUK, S. H. Effect of soil properties on erosion by wash and splash. **Earth Surface Processes**, n. 4, 1979. p. 241-255.

MAGOSSI, L. R. & **BONACELLA**, P. H. Poluiçao das aguas - 3rd ed. - São Paulo : Moderna, 2013. - (challenges collection).

MARQUES, J. S. Geomorphological science. In: GUERRA, A. J. T. ; CUNHA, S. B. (Org.). Geomorfologia: **uma actualizaçao** de **bases e conceitos**, 4. Ed. Rio de Janeiro: Bertrand Brasil, 2001. p. 23-50.

MARCONI, M.A. & **LAKATOS,** E. M. Técnicas de Pesquisa: **planejamento e execuçâo de pesquisas, amostragens e técnicas de pesquisa, elabor elaboraçãoçâo, anâlise e interpretaçâo de dados,** 7ª ed. - 7ª reimp. - Sao Paulo : Atlas, 2013.

MARTINS, S. V. Recovery of riparian forests. **Aprenda Fâcil,** ed. Viçosa, MG. 2nd edition, 2007. 255 pp.

MOTA, S. **Introduçâo a Engenharia Ambiental,** 4th ed. Rio de Janeiro: Expressao Gràfica, 2010.

MOURA, D. G. Working with projects: **planning and managing educational projects** / Dàcio G. Moura, Eduardo F. Barbosa. 8th ed. - Petrópolis, RJ : Vozes, 2013.

NASCIMENTO NETO, P. Residuos sólidos urbanos: **perspectivas de gestão intermunicipal em regiôes metropolitanas.** Sao Paulo: Atlas, 2013.

NASCIMENTO, J. N. S. et al., **Geographic and Climatic Transformations of Deforestation in Itaituba.** Anais Artigos, III SIMPÓSIO de estudos e Pesquisas em Ciências Ambientais na Amazônia, v.3, ISSN: 23167637 - 2014.

NASCIMENTO, J. N. S. Geographical and climatic transformations of deforestation in Itaituba. In'**ALMEIDA,** Djalmira de Sà (Org.) **Producing academic texts**: basic sanitation, health, culture, society and sustainable environment in Tapajós. Curitiba - PR, 2015, p : 151.

NASCIMENTO, J. N. S. Amazônia do século XVI, uma abordagem histórica, econômica e social, rumo ao século XXI. In'**ALMEIDA,** Djalmira de Sà (Org.) **Producing academic texts**: basic sanitation, health, culture, society and sustainable environment in Tapajós. Curitiba - PR, 2015, p : 149 - 155.

OLIVEIRA, J. Z. M. **Residuos Sólidos e Educaçâo Ambiental numa perspectiva da análisis geogràfica.** Manaus: Editora da Universidade Federal do Amazonas, 2009. p. 243-258.

OLIVEIRA, M. S. P. et al., **Açai (Euterpe oleracea Mart.). EMBRAPA,** s/r.

PHILIPPI JR. A., **GALVÂO JR. & CASTRO,** A. Gestao do Saneamento Bàsico: **abastecimento de água e esgotamento sanitrio.** Barueri SP: Manole 2012, (environmental collection).

PIRETE, L. M. & et al., Article Evaluation of the final urban solid waste disposal area in the municipality of Araguai using the waste landfill quality index - IQR. Revista Agrogeoambiental, special ed. no. 2, Federal Institute of Education, Science and Technology South of Minas Gerais, 2014.

PORTAL AMAZÔNIA, 2010.

PREGELLE, D. R. et al, **Recovering springs in cerrado areas,** EMBRAPA gado de corte, Campo Grande, Brazil. II International Symposium on Tropical Savannas, ParlaMundi, Brasilia, DF. 2008.

Manual for the recovery of degraded areas in the state of São Paulo riparian forests in the interior of São Paulo, Guaratinguetà/SP, 2006.

RESENDE, S. C. **& HELLER,** L. O Saneamento no Brasil: **Politicas e Interfaces,** 2ª ed. ver. E ampl Belo Horizonte: ed. UFMG, 2008.

ROCHA, E. E. R. B., **Participatory research and its developments - experiences in popular organizations.** Proceedings of the II Brazilian Congress of University Extension, Belo Horizonte, 2004.

RODRIGUES, L. N., **Estatistica experimental:** Castanhal, 2014.

RODRIGUES, T. E., **VALENTE**, M. A. Aspectos Gerais Sobre Adubaçao e Calagem, in' Recomendações de Adubaçao e Calagem para o Estado do Parà, CRAVO, M. S., VIÉGAS, I. J. M., BEASIL, E. C. EMBRAPA, 2007.

RONDON, E. V. **Biomass Production and Growth of *Schizolobium Amazonicum*** (Huber) Ducke **Trees** under Different Spacing in the Forest Region, Viçosa - MG, 2002.

ROSSA, U. B. et al., Slow-release fertilization in the growth of paricà seedlings in nurseries, Curitiba, 2013.

SACRAMENTA, D. M O. & **COSTA,** B. P. The agrarian question and migration in the Brazilian Amazon: **the case of the canoas settlement in the municipality of Presidente**

Figueiredo/AM. s/r.

SALMAN, A. K. D. et al. Native tree species of the Brazilian Western Amazon with potential for pasture afforestation, EMBRAPA 2008.

SANTOS FILHO, B. G. dos et al., **Nutritional Behavior of Tree Species Used in the Reforestation of Degraded Areas Under the Impact of Oil Exploration in the Urucu Region, Municipality of Coari, AM,** scientific note, 2007.

SCHONS, D. C. et al., Article Monitoring the quality of the waters of the Ouro River, in Ouro Verde do Oeste - PR: Toxicological Analyses. Revista Agrogeoambiental, special ed. no. 2, Federal Institute of Education, Science and Technology South of Minas Gerais, 2014.

SOUSA, L. H. S. de & et al., **Consumption X Waste Generation.** Anais Artigos, III SIMPÓSIO de estudos e Pesquisas em Ciências Ambientais na Amazônia, v.3, ISSN: 23167637 - 2014.

SOUSA, L. H. S. et al, A influência das queimadas e suas consequências para o Solo e o Meio Ambiente na regiao Sudoeste do Parâ. In'**ALMEIDA,** Djalmira de Sà (Org.) **Producing academic texts**: basic sanitation, health, culture, society and sustainable environment in Tapajós. Curitiba - PR, 2015, p : 35.

SOUZA, A. S., Article Species of the urban fauna of Marabà and Parauapebas: knowing to preserve. Revista Agrogeoambiental, special ed. no. 2, Federal Institute of Education, Science and Technology South of Minas Gerais, 2014.

SOUZA, C. R. et al., Technical Circular 18. EMBRAPA, (2003).

TREVISAN, E. C. **Protected and recovered springs** - Curitiba: SEMA, 2010. P. 24.

UZZO, R. P. Doctoral Thesis: Physiological and Anatomical Response of Açaizeiro and Royal Australian Palm to Shading. Sao Paulo, 2008.

VIDAL, J. P. Municipal public management: a transformative instrument for social development in Parà. Third margin Amazon. Outras expressões v. 1, n 1,Sao Paulo - Jun/2011. p. 15-33.

VIEIRA, C. R. & **WEBER,** O. L. S. Organic Compounds in the Initial Growth of Paricà Seedlings - IV Brazilian Congress of Environmental Management - Salvador/BA, 2013.

ABOUT THE AUTHOR

PhD student in Human Geography (USP 2017), Master in Rural Development and Management of Agri-Food Enterprises IFPA (29/01/2016), Specialist in Cultural Studies of the Amazon - UFPA (2006/2007), Specialist in School Pedagogy, with specialization in Administration, Supervision and Orientation - IBPEX (2005/2006), degree in Pedagogy under special regime - Full Degree from Vale do Acaraù State University (2005). Degree in History from Faculdade de Itaituba (2007) and Technical Degree in Agriculture from Colégio Agricola Caetano Costa (1983). He is currently working as a Technical and Technological Teacher at IFPA - Campus Itaituba. His research focuses on the following themes: Primary Education, EJA and Secondary Education, Local History, Agroecology, Agroforestry Systems, Recovering Degraded Areas and the Tapajós Hydrographic Basins.

Printed by Books on Demand GmbH, Norderstedt / Germany